Analytical Methods in Radiative Transfer

Wiley Series in Atmospheric Physics and Remote Sensing

Series Editor: Alexander Kokhanovsky

Wendisch, M. / Brenguier, J.-L. (eds.)

Airborne Measurements for Environmental Research Methods and Instruments

2013

Coakley Jr., J. A. / Yang, P.

Atmospheric Radiation: A Primer with Illustrative Solutions

2014

Stamnes, K. / Stamnes, J. J.

Radiative Transfer in Coupled Environmental Systems An Introduction to Forward and Inverse Modeling

2015

Tomasi, C. / Fuzzi, S. / Kokhanovsky, A.

Atmospheric Aerosols Life Cycles and Effects on Air Quality and Climate

2016

Weng, F.

Passive Microwave Remote Sensing of the Earth for Meteorological Applications

2017

North, G. R. / Kim, K.-Y.

Energy Balance Climate Models

2017

Kokhanovsky, A. / Natraj, V. / Efremenko, D.

Analytical Methods in Radiative Transfer

2025

Analytical Methods in Radiative Transfer

Alexander Kokhanovsky

Philipps-Universität Marburg, Department of Geography, Laboratory for Climatology and Remote Sensing, Germany

Vijay Natraj

Jet Propulsion Laboratory, California Institute of Technology, USA

Dmitry Efremenko

Remote Sensing Technology Institute (IMF), German Aerospace Center (DLR), Germany

WILEY-VCH

Authors

Dr. Alexander Kokhanovsky
Philipps-Universität Marburg
Department of Geography
Laboratory for Climatology and
Remote Sensing
F|14, Deutschhausstr. 12
35032 Marburg, Germany

Dr. Vijay Natraj
Jet Propulsion Laboratory
California Institute of Technology
4800 Oak Grove Drive
Pasadena
91109 CA, USA

Dr. Dmitry Efremenko
DLR
Münchener Str. 20
82234 Weßling, Germany

A book of the Wiley Series in
Atmospheric Physics and
Remote Sensing

The Series Editor

Dr. Alexander Kokhanovsky
Philipps-Universität Marburg
Department of Geography
Laboratory for Climatology and
Remote Sensing, Germany

Cover Design: Wiley
Cover Image: © NASA (public domain)

Library of Congress Card No.: applied for

British Library Cataloguing-in-Publication Data
A catalogue record for this book is available from the British Library.

Bibliographic information published by the Deutsche Nationalbibliothek
The Deutsche Nationalbibliothek lists this publication in the Deutsche Nationalbibliografie; detailed bibliographic data are available on the Internet at http://dnb.d-nb.de.

Print ISBN 9783527411436
ePDF ISBN 9783527698943
ePub ISBN 9783527698974
oBook ISBN 9783527698950

Typesetting Lumina Datamatics
Printing CPI Group (UK) Ltd, Croydon, CR04YY
C9783527411436_220825

The manufacturer's authorized representative according to the EU General Product Safety Regulation is Wiley-VCH GmbH, Boschstr. 12, 69469 Weinheim, Germany, e-mail: Product_Safety@wiley.com.

Dedication

We dedicate this book to our families for their patience and support.

Contents

Preface

This book serves as a foundational guide to analytical techniques in radiative transfer theory, focusing on solving the radiative transfer equation (RTE). The RTE is crucial across diverse research fields, including atmospheric optics, astrophysics, and nuclear reactor design. It plays a pivotal role in interpreting the interaction of radiation with heterogeneous media such as clouds, snow, ice sheets, atmospheric aerosols, and oceans, among other applications. Typically, the RTE is addressed through numerical solutions, supported by various robust and swift radiative transfer codes, many of which can be found online.[1]

Solving the RTE enables detailed analysis of radiative transfer problems, including the determination of the spectral and angular distribution of radiance (or light intensity) across the boundaries and within the body of turbid media. A significant aspect of this field is the employment of analytical models, especially useful under conditions of small or large parameter values, such as the likelihood of photon absorption, optical thickness, or the average cosine of scattering angle. These models are not just grounded in sophisticated physical concepts but also facilitate a profound comprehension of radiative transfer theory, showcasing its elegance and significance.

Analytical models hold immense importance as they offer a balanced mix of simplicity and accuracy, making them indispensable for situations where exact solutions are infeasible due to computational constraints or a lack of detailed information about the medium. They enable researchers and practitioners to predict and analyze radiative characteristics efficiently, fostering advancements in both theoretical understanding and practical applications.

The significance of analytical models extends beyond academic interest, influencing practical decision-making in environmental monitoring, climate change predictions, and even in medical imaging technologies. By enabling quicker and more accessible analyses, these models facilitate the timely assessment of critical phenomena, such as the assessment of sunlight penetration in oceans affecting marine ecosystems or the evaluation of atmospheric conditions for climate modeling. Their adaptability and efficiency make them invaluable tools in the quest to understand and mitigate complex environmental challenges.

---

1 https://en.wikipedia.org/wiki/Atmospheric radiative transfer codes

As the volume of data collected from satellites continues to expand exponentially, there is a pressing need for more rapid techniques to interpret this vast influx of information. This challenge has spurred researchers to explore innovative solutions, often revisiting and adapting analytical models developed in the 20th century to meet contemporary demands. By integrating these traditional models with cutting-edge machine learning algorithms, researchers are achieving breakthroughs that not only match the accuracy of more time-intensive approaches but also dramatically reduce processing times, often by several orders of magnitude.

By emphasizing these models, this book aims to provide an exhaustive overview of analytical RTE solutions, aspiring to motivate future scientists to pioneer in utilizing the RTE for solving both direct and inverse radiative transfer problems. Through this endeavor, we hope to contribute to the ongoing development and application of radiative transfer theory, underpinning new discoveries and technological innovations.

We hope that newcomers embarking on their journey into the field of radiative transfer will find this book engaging and enlightening. It is designed to provide simple expressions that demystify the subject, enabling readers to grasp its core physical principles rather than getting bogged down in complex mathematical formulas. Our goal is to make the foundational concepts of radiative transfer accessible and comprehensible, offering a practical approach that enhances intuitive understanding. This approach allows those just starting out not only appreciate the elegance and relevance of radiative transfer but also apply its principles effectively across a variety of scientific and engineering contexts. By focusing on the physical rather than purely mathematical aspects, we aim to inspire broader application of radiative transfer theory.

Marburg, Pasadena, Herrsching
July 2025

A. Kokhanovsky, V. Natraj, D. Efremenko

Acknowledgments

A portion of the research involved in writing this book was carried out at the Jet Propulsion Laboratory, California Institute of Technology, under a contract with the National Aeronautics and Space Administration (80NM0018D0004).

1

Introduction

1.1 Historical Background

Perhaps the theory of radiative transfer originated at the dawn of human civilization at the moment when a person thought about why the Sun warms. The gradual realization of what radiation is eventually led to the birth of quantum mechanics. Probably among the three types of heat transfer (thermal conduction, convection, and radiation), it was the phenomenon of radiation that was the most mysterious for ancient researchers. Hippocrates of Kos, an ancient Greek physician and philosopher (460–370 BC), said, "Light is a mystery to us. We do not understand what it is, or how it can be. We know only that it exists, and that it is beautiful."

Several early researchers made significant contributions to our understanding of radiative transfer. Their work continues to influence research in this field even today. Johann Lambert (1728–1777) studied how the incident light reflects from a surface depending on the angle of incidence. He found that the radiant intensity or luminous intensity observed from an ideal diffusely reflecting surface is directly proportional to the cosine of the angle θ between the observer's line of sight and the surface normal (Lambert, 1760):

$$I = I_0 \cos\theta, \tag{1.1}$$

where I is the intensity of the reflected light, and I_0 is the intensity of the incident light. A surface that obeys Lambert's law is said to be Lambertian and exhibits Lambertian reflectance. Such a surface has the same radiance (luminance) when viewed from any angle. This means, for example, that to the human eye it has the same apparent brightness. It has the same radiance because, although the emitted power from a given area element is reduced by the cosine of the emission angle, the solid angle, subtended by surface visible to the viewer, is reduced by the very same amount. Because the ratio between power and solid angle is constant, radiance (power per unit solid angle per unit projected source area) stays the same. Despite its simplicity, the Lambert cosine law has found many practical applications, such as lighting design, solar energy collection, and computer graphics. In remote sensing, this law is used to model the reflection of electromagnetic radiation as it interacts with the Earth's surfaces. Lambert introduced two axioms of photometry: light travels in a straight line in a uniform medium and rays that cross do not interact. It was also shown that light intensity decays exponentially in an absorbing medium. All results of Lambert

Analytical Methods in Radiative Transfer, First Edition. Alexander Kokhanovsky, Vijay Natraj and Dmitry Efremenko.

were summarized in his book *Photometria,* which is considered a landmark in the field of photometry, as it introduced several important concepts and techniques that are still used today.

Pierre-Simon Laplace (1749–1827) made significant contributions to the theory of radiative transfer. Laplace was interested in understanding the nature of energy transfer in the Earth's atmosphere. He was one of the first scientists who realized the important role of radiative transfer in the energy balance of the Earth system. Laplace developed the theory of probabilities for describing the processes occurring in nature. He formulated what became known as the central limit theorem and proved that the distribution of errors in large data samples from astronomical observations can be approximated by a normal distribution. The description of the behavior of radiation, as it interacts with matter using probabilities and statistics, appeared to be very useful and became the fundamental concept of radiative transfer.

One of the first scientists during the late modern period who investigated the properties of radiation and conducted a quantitative study of radiative transfer was John Tyndall (1820–1893). One of his experiments involved measuring the infrared emission of a platinum filament and observing the corresponding color of the filament. A platinum wire was placed inside a glass tube filled with nitrogen gas. An electric current passed through the wire heated it, causing it to emit radiation. Tyndall used a thermopile to measure the intensity of the radiation emitted by the wire. He also observed that as the wire was heated, it first appeared red, then orange, yellow, white, and finally blue-white. This sequence of colors is referred to as the Tyndall effect. Tyndall measured the intensity of the radiation emitted by the wire at different wavelengths using a prism and a spectroscope. He found that as the wire was heated, the intensity of the radiation increased at shorter wavelengths, corresponding to the blue end of the spectrum. This experiment demonstrated the relationship between the color of a heated object and the wavelengths of radiation emitted by the object. It also provided important insights into the properties of infrared radiation, which is invisible to the human eye but plays a crucial role in many physical processes, such as heat transfer and climate change.

Based on experiments of Tyndall, Josef Stefan (1835–1893) found the proportionality of the intensity to the fourth power of the absolute temperature (see Figure 1.1). Later, Ludwig Boltzmann explained this result theoretically, what is now referred to as the Stefan–Boltzmann law (Boltzmann, 1884). It states that the total radiant heat power E emitted from the surface of a blackbody across all wavelengths is proportional to the fourth power of its absolute temperature T:

$$E = \sigma T^4, \tag{1.2}$$

where $\sigma \approx 5.67 \cdot 10^{-8}\mathrm{W/(m^2K^4)}$ is the Stefan–Boltzmann constant. The discovery of Stefan–Boltzmann law made it possible to estimate quite accurately the temperature of the Sun's surface (Stefan, 1879). Now we know that any object emits radiation. It is even possible to visualize radiation using thermal imagers. As shown in Figure 1.2, faces and hands are the warmest "objects" and hence have the most intensive emission.

In 1900, Max Planck formulated his law as part of his efforts to solve the problem of blackbody radiation, which had puzzled physicists for decades. Planck's law is based on the

Um die Wärmemenge zu erhalten, welche dieselbe Fläche bei der Temperatur 100° ausstrahlt, hat man diese Zahl mit a^{100} $= 2{\cdot}146$ zu multipliciren und erhält 1·6644. Die Differenz der beiden 0·889 gibt die Wärmemenge, welche durch die Einheit der Oberfläche einer Glaskugel von 100° Temperatur in einer Glashülle von 0° per Minute abgegeben wird. Es stimmt diese Zahl mit der oben mit Zuziehung der Correction wegen der Wärmeleitung berechneten gut überein.

Wählt man für das Gesetz der Strahlung die Formel der vierten Potenzen der absoluten Temperaturen, so ist

$$H_1 = AT_1^4, \quad H_2 = AT_2^4$$

Figure 1.1 The first reference to the fourth degree law – an excerpt from the article of Josef Stefan (Stefan, 1879). Translation: "To obtain the amount of heat which the same surface radiates at a temperature of 100°, this number must be multiplied by $a^{100} = 2.146$ and 1.6644 is obtained. The difference between the two 0.889 gives the amount of heat, which is given off per minute by the unit surface of a glass sphere at a temperature of 100° in a glass envelope at 0°. This number agrees well with the one calculated earlier with the correction for heat conduction. If one chooses the formula of the fourth power of the absolute temperatures for the law of radiation, then $H_1 = AT_1^4$, $H_2 = AT_2^4$."

Figure 1.2 Example of an infrared image from a thermal imager. The skin has the highest temperature and hence the highest emission.

assumption that electromagnetic radiation is quantized, or composed of discrete packets of energy, rather than being continuous. According to Planck's law, the spectral density of energy emitted by a blackbody is proportional to the frequency of the radiation, and it is also affected by the temperature of the blackbody. This was a revolutionary concept in physics, which required a revision of most fundamental concepts. The mathematical expression for the Planck's law is:

$$B(\lambda, T) = \frac{2hc^2}{\lambda^5} \frac{1}{e^{\frac{hc}{\lambda kT}} - 1}, \tag{1.3}$$

where B is the spectral radiance, λ is the wavelength of the radiation, h is the Planck's constant, c is the speed of light, and k is the Boltzmann constant. It can be shown by integrating Eq. (1.3) across all wavelengths that the Stephan–Boltzmann law is a consequence of the Planck's law.

Wien's displacement law reveals that the peak wavelength of blackbody radiation shifts in a manner inversely proportional to the temperature, indicating that with higher temperatures, the radiation peak moves toward shorter wavelengths, namely

$$\lambda_{\max} T = b, \tag{1.4}$$

where $\lambda_{\max}$ is the wavelength at which the radiation curve has a peak, while $b \approx 2898\ \mu\text{m}\cdot\text{K}$ is the Wien's displacement constant. This fact is derived directly from the Planck radiation law. However, Wilhelm Wien (1864–1928) formulated his law years before Max Planck introduced his more comprehensive formula. In his derivation, Wien was based on the so-called thermodynamic argument.

Wien considered a cavity filled with electromagnetic radiation in thermal equilibrium at a certain temperature. He then imagined adiabatically compressing or expanding the cavity. According to the laws of thermodynamics, such an adiabatic process would change the frequency of the photons inside the cavity (due to the Doppler effect and the change in boundary conditions affecting the standing wave modes of the cavity), but it would not lead to any exchange of heat with the surroundings. The adiabatic compression or expansion of the cavity causes a proportional shift in the frequency (or wavelength) of the radiation inside. Because this process is adiabatic, the frequency shift is directly related to a change in the temperature of the radiation. The functional form of the radiation spectrum must therefore change in a specific way to maintain thermal equilibrium at a new temperature. Wien reasoned that the spectral distribution of radiation in the cavity must scale in a specific way with temperature to satisfy the requirements of thermodynamics. This led him to conclude that the product of the peak wavelength of the radiation and the temperature of the blackbody must be constant. This relationship is now known as Wien's displacement law. Peak wavelengths and temperatures for some objects are shown in Table 1.1.

Finally, in the 1860s, James Clerk Maxwell developed a theory that described the behavior of electric and magnetic fields and their interactions with matter. Maxwell's theory is considered one of the most significant achievements in the history of physics and laid the foundation for much of modern electromagnetism and telecommunications. Maxwell's theory is composed of four partial differential equations that relate the electric and magnetic fields to their sources, which can be electric charges or currents[1]:

---

1 Despite having six unknown components (three each for the electric and magnetic fields), the four Maxwell equations yield eight scalar equations. It has been shown that if a system meets the conditions of Faraday's law and Ampere's law, it will also satisfy both Gauss's laws, provided the system's initial conditions do so and under the assumptions of charge conservation and the absence of magnetic monopoles (Rosen, 1980).

Table 1.1 Objects, their temperatures, and peak wavelengths of thermal emission.

Object	T (K)	λ_{max} (μm)
Cosmic microwave background	2.725	1062.88
Neptune	59	49.15
Earth	288	10.06
Human body	310	9.35
Cool red star (Proxima Centauri)	3050	0.95
Venus	737	3.93
Sun	5778	0.50
Hot blue star (Rigel)	11000	0.26
White dwarf (Sirius B)	25200	0.11

- Gauss's Law for Electricity:

$$\nabla \cdot \mathbf{E} = \frac{\rho}{\epsilon_0}, \tag{1.5}$$

where **E** is the electric field, ρ is the electric charge density, and ϵ_0 is the permittivity of free space. This equation states that the electric flux out of any closed surface is proportional to the charge enclosed.
- Gauss's Law for Magnetism:

$$\nabla \cdot \mathbf{B} = 0, \tag{1.6}$$

where **B** is the magnetic field. This equation implies that magnetic monopoles do not exist, as the net magnetic flux out of any closed surface is zero.
- Faraday's Law of Electromagnetic Induction:

$$\nabla \times \mathbf{E} = -\frac{\partial \mathbf{B}}{\partial t}. \tag{1.7}$$

Here t represents time. This law indicates that a changing magnetic field induces an electric field.
- Ampére's Law with Maxwell's Addition:

$$\nabla \times \mathbf{B} = \mu_0 \mathbf{J} + \mu_0 \epsilon_0 \frac{\partial \mathbf{E}}{\partial t}, \tag{1.8}$$

where μ_0 is the permeability of free space and **J** is the current density. This law shows that magnetic fields can be generated by electric currents and by changing with time electric fields.

Maxwell's equations are significant because they unify the fields of electricity and magnetism, which were previously considered to be separate phenomena. They also predict the existence of electromagnetic waves, which are disturbances in the electric and magnetic fields that propagate through space at the speed of light. In principle, Maxwell's equations give a theoretical framework for describing the radiative transfer phenomena from the basic principles. However, this approach appeared to be very complex from the computational point of view. Therefore, its applications are very limited up to now. In this regard, the radiative transfer theory evolved as a separate discipline in physics.

1.2 What Is Radiative Transfer About and What Is It Good For?

Radiative transfer refers to the transfer of electromagnetic radiation (such as light) as it interacts with matter in a medium (such as a gas, liquid, or solid). It is a fundamental process that occurs in many natural and artificial systems, including the Earth's atmosphere, hydrosphere, cryosphere, oceanic and other terrestrial surfaces, stars and galaxies, and many industrial processes. The study of radiative transfer involves understanding how radiation is emitted, absorbed, and scattered by matter, and how it is influenced by factors such as temperature, pressure, and composition. Radiative transfer has many applications in science and engineering. In atmospheric science, radiative transfer models are used to understand the behavior of radiation in the Earth's atmosphere and to study the effects of atmospheric composition and climate change (Liou, 2002b). In astrophysics, radiative transfer models are used to study the properties of stars and galaxies and to understand the evolution of the universe (Rybicki and Lightman, 1985). In industrial processes, radiative transfer models are used to design and optimize energy-efficient systems for heating, cooling, and lighting.

Radiative transfer is also important in many fields of engineering, including aerospace, optics, and materials science. For example, in aerospace engineering, radiative transfer models are used to understand the behavior of radiation in space and to design spacecraft that can withstand extreme temperatures and radiation levels. In optics, radiative transfer models are used to design and optimize optical devices such as lenses and filters and to understand the behavior of light in complex media such as biological tissues. In materials science, radiative transfer models are used to study the thermal properties of materials and to design materials with specific optical and thermal properties (Chen et al., 2016). Table 1.2 summarizes the applications of radiative transfer theory.

Table 1.2 Applications of radiative transfer theory.

Field	Context	Applications
Climate Science and Meteorology	Understanding the Earth's atmosphere's interaction with solar radiation, influencing climate patterns, weather forecasting, and climate change studies.	Global climate models, weather prediction models
Astronomy and Astrophysics	Interpreting light from stars and galaxies to deduce compositions, temperatures, motions, and ages; studying star formation and galaxy structures.	Star and galaxy analysis, cosmic background studies

Remote Sensing and Satellite Imaging	Interpreting Earth observation satellite data for mapping surfaces, monitoring vegetation, assessing pollution, and observing planetary bodies.	Land and ocean mapping, environmental monitoring
Optical Engineering	Designing optical devices by optimizing light propagation through components like lenses and filters.	Telescope, camera, and laser design
Biomedical Applications	Applications in medical imaging technologies (e.g. magnetic resonance imaging, computer tomography scans) and therapeutic interventions (photodynamic therapy, laser surgery) by understanding light-tissue interactions.	Medical diagnostics and treatments
Nuclear Reactor Design	Analyzing neutron flux distributions in reactor designs, where radiative transfer principles apply analogously to neutron propagation.	Safety and efficiency optimization in nuclear reactors

1.3 Phenomenological Radiative Transfer

The study of radiative transfer can be approached in two different ways, namely, the phenomenological approach and the microphysical approach.

Radiative transfer theory was initially developed as a phenomenological approach that relied heavily on intuition and experience-based shortcuts (Kravtsov and Apresyan, 1996). It focuses on describing how radiation interacts with matter at a macroscopic level, using empirical data and observations to formulate models and theories. The main assumptions of this approach are the following:

1. The use of the ray approximation, which assumes that radiation propagates in the form of rays and is valid in the limit of geometrical optics;
2. The assumption that rays arriving at a given point are incoherent and can be algebraically summed up without regard for phases or interference terms;
3. The use of time- and space-averaged squared field characteristics instead of local or instantaneous values;
4. Principle of ergodicity, which means that averaging over time for a given realization of a random scattering process is equivalent to ensemble averaging;
5. The medium is assumed to consist of randomly distributed scatterers, and the interaction between the ray and the medium is statistically independent of the outcome of subsequent scattering events, which is known as a Markov process. The particles are randomly oriented.

While the phenomenological approach is advantageous due to its simplicity and efficient numerical models for simulating scattered radiance, it is limited in its accuracy when the wave properties of light are important. In particular, this approach cannot accurately predict coherent backscattering (Barabanenkov et al., 1991), i.e. constructive interference of multiple scattered paths that have the same length but are reversed in direction. Coherent backscattering results in an enhancement of radiance in the backscattering direction observed in an extremely sharp cone around that direction.

1.4 Microphysical Approach

In the microphysical approach, radiative transfer is viewed as the propagation of electromagnetic waves in a discrete medium comprising a set of scatterers. The radiative transfer theory is considered as a consequence of Maxwell's equations (Ishimaru, 1978). As a matter of fact, in the microphysical approach, it is essential to determine conditions that allow us to define the quantities used in the radiative transfer theory and derive the radiative transfer equation using Maxwell's equation as a starting point. A comprehensive review of discussions on microphysical and phenomenological approaches can be found in Mishchenko (2013). Rigorous derivations of the microphysical approach are outside the scope of this book, but can be found elsewhere (Mishchenko, 2008; Doicu and Mishchenko, 2018, 2019). Here, we provide a brief outline of the main ideas.

The medium is assumed to consist of discrete particles; each particle scatters the electromagnetic wave, and in the case of a set of particles, the radiation is scattered multiple times before it is measured by the detector at the observation point r'. The total electric field $\mathbf{E}$ at r' is the superposition of the respective incident $\mathbf{E}^{\mathrm{inc}}$ and N partial scattered fields $\mathbf{E}_i^{\mathrm{sca}}$:

$$\mathbf{E}(r', t) = \mathbf{E}^{\mathrm{inc}}(r', t) + \sum_{i=1}^{N} \mathbf{E}_i^{\mathrm{sca}}(r', t). \tag{1.9}$$

Here $\mathbf{E}_i^{\mathrm{sca}}$ corresponds to the wave, scattered by j-th particle. For the magnetic field $\mathbf{H}(r', t)$, the equation looks analogical. To proceed with mathematical derivations, the microphysical approach usually assumes that the particles are randomly and uniformly distributed, and their positions are uncorrelated. While an ensemble average over all position realizations would typically be required, assuming the property of ergodicity allows for time averaging instead. As the scattered wave travels a sufficient distance from the scattering particle, it becomes spherical. The far-field approximation assumes that all particles are in the far-field of each other, thereby making the medium sparse.

Furthermore, the i-th particle scatters waves that may have previously been scattered by other particles. Thus, the field that strikes the i-th particle reads as

$$\mathbf{E}(r_i, t) = \mathbf{E}^{\mathrm{inc}}(r_i, t) + \sum_{\substack{j=1 \\ i \neq j}}^{N} \mathbf{E}_j^{\mathrm{sca}}(r_i, t). \tag{1.10}$$

In each turn, $\mathbf{E}_j^{\mathrm{sca}}$ is a result of scattering of the striking wave $\mathbf{E}\left(r_j, t\right)$. The resulting electric and magnetic fields at the observation point can be expressed as order-of-scattering series (Foldy, 1945; Lax, 1951). There the resulting field is represented as a sum of fields being scattered by all possible sequences of particles. However, Mishchenko (2014) warned

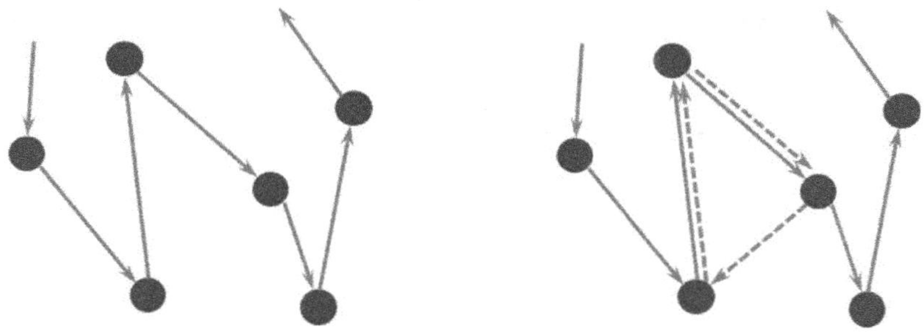

Figure 1.3 Scattering sequences: (left) self-avoiding scattering path, which is allowed by the Twersky approximation; (right) non-self-avoiding scattering path, which is not allowed by the Twersky approximation. The dashed line marks a closed loop in the trajectory.

against perceiving this scheme as a physical picture and not a mathematical construction since all mutual excitations occur simultaneously.

To evaluate **E** and **H** at the observation point, contributions from all sequences of scattering have to be summed up. To perform analytical summation, the Twersky approximation (Twersky, 1964) is used. Specifically, it is assumed that the radiation follows self-avoiding scattering paths, where particles appear only once. The examples of self-avoiding and non-self-avoiding trajectories are shown in Figure 1.3. In the Twersky approximation, the coherent component of radiation is neglected, and the contributions from different scattering sequences are performed disregarding the phases of waves.

The Poynting vector is a fundamental concept in electromagnetism that describes the flow of energy in electromagnetic waves. It is defined as:

$$\mathbf{S} = \mathbf{E} \times \mathbf{H}, \tag{1.11}$$

In the context of radiative transfer, the Poynting vector describes the flow of energy through a medium that scatters or absorbs radiation. As electromagnetic waves propagate through the medium, they transfer energy to the particles in the medium. The scattered or absorbed radiation can be quantified by integrating the Poynting vector over a surface enclosing the medium. This integration yields the total power that the electromagnetic field conveys to the medium, effectively capturing the dynamic interaction between the electromagnetic waves and the medium at a macroscopic level. Since the Poynting vector **S** oscillates with a period of about $\sim 10^{15}$ seconds at optical frequencies and optical detectors are characterized by a time constant, which is much larger than such a period, the time-averaged value of the Poynting vector is measured. To compute a time-averaged Poynting vector, it is convenient to consider **E** and **H** as complex fields. In this case, the time-averaged Poynting vector can be represented as

$$\langle \mathbf{S} \rangle = \frac{1}{2} \mathrm{Re}\, \{\mathbf{E} \times \mathbf{H}^*\}, \tag{1.12}$$

where "*" denotes complex conjugate.

When considering the scattering of electromagnetic waves by a collection of particles, the same particle can appear in both the **E** and **H** scattering sequences, since the electric and magnetic fields are scattered independently. To address this, the ladder approximation is employed, which assumes that the orders of common particles are the same in both

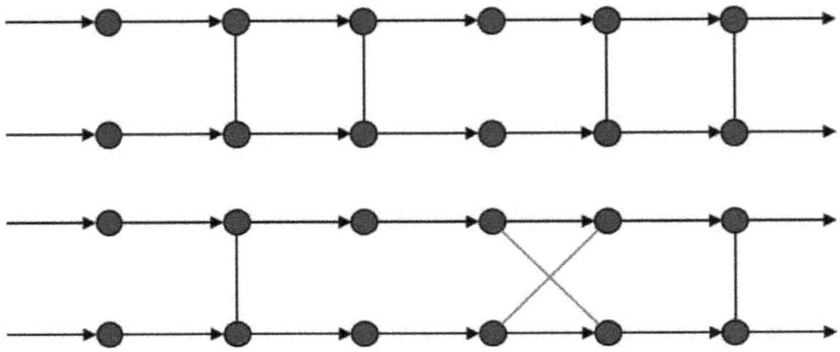

Figure 1.4 Feynman diagrams: (top) in the case of the ladder approximation, (bottom) with crossing connectors.

sequences. This approximation is valid for systems with a large number of particles, similar to the Twersky approximation. By making these assumptions, it becomes possible to perform an analytical summation of the scattering sequences, leading to the derivation of the radiative transfer equation. This is an important result, as it provides a means of quantifying the transport of energy through a medium in terms of the Poynting vector.

Since **E** and **H** are scattered independently by an ensemble of particles, there may be common particles in the **E** and **H** scattering sequences. Let us consider these particles on the Feynman diagrams (Figure 1.4), wherein the arrows indicate the order of scattering while the vertexes connect the same particles. In the so-called ladder approximation, the orders of common particles are the same in both sequences for **E** and **H**. By making these assumptions, it becomes possible to perform an analytical summation of the scattering sequences, leading to the derivation of the radiative transfer equation. For mathematical details, interested readers are encouraged to refer to recent reviews (Doicu and Mishchenko, 2018, 2019).

1.5 Atmospheric Remote Sensing

Remote sensing is the collection of information about an object or environment without direct physical contact. This is achieved through the use of various sensors that capture and measure electromagnetic radiation emitted or reflected by the target. In particular, atmospheric remote sensing is a technique for observing and measuring the physical properties of the Earth's atmosphere from a distance using sensors and instruments mounted on satellites or airplanes. It has a wide range of applications, including weather forecasting, climate research, air quality monitoring, and disaster management.

Satellite remote sensing has revolutionized our ability to observe Earth, providing a comprehensive view that was once impossible. This technology has yielded critical insights into the physical and chemical processes of the atmosphere, impacting areas such as climate change, air quality, and weather forecasting. Tools like NASA Worldview[2] and Google Earth[3] enhance our understanding by offering real-time environmental data and detailed

2 https://worldview.earthdata.nasa.gov/

3 https://earth.google.com/web/

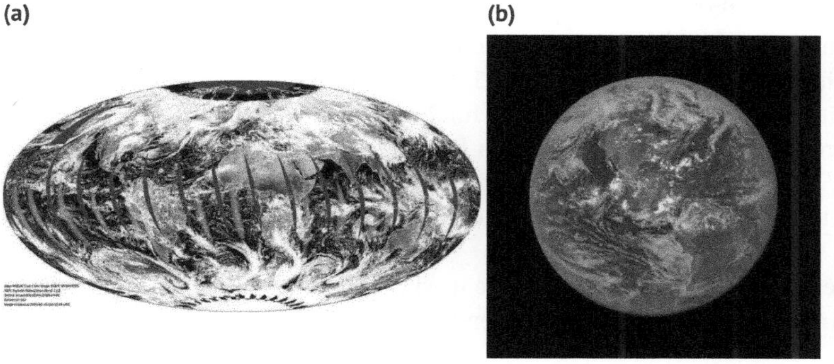

Figure 1.5 Example of an image (originally in color; rendered in grayscale for print) provided by (a) MODIS Aqua on February 08, 2023. (b) DSCOVR EPIC from the Lagrange point. *Source*: (a) Credits: (a) NASA / Public domain (b) Marshall Sutton / NASA / Public domain.

global views, respectively, making the vast data collected by satellites accessible and usable for experts and the public alike. Additionally, the DSCOVR EPIC camera,[4] positioned at the Lagrangian Point 1 (L1),[5] offers unique perspectives on the Earth's atmosphere and surface, enhancing our understanding of environmental and atmospheric dynamics. As illustrated in Figure 1.5, satellites are indispensable for atmospheric remote sensing, enabling consistent, systematic global coverage. Satellite-based remote sensing of the atmosphere has provided significant insights into the physical and chemical processes occurring in the atmosphere, including those related to climate change, air quality, and weather forecasting.

Satellites used in atmospheric remote sensing are equipped with sensors that measure the interaction of the atmosphere with electromagnetic radiation in different spectral regions. The sensors can measure UV, visible, infrared, and microwave radiation, among other wavelengths. One of the key advantages of satellite-based remote sensing is the ability to collect data over large areas, including remote and inaccessible regions such as the polar regions or oceans. This data can be used to better understand global processes, such as the transport of pollutants or the movement of weather systems. Satellites also provide temporal coverage, allowing for the monitoring of changes in the atmosphere over time. This is particularly important for tracking long-term changes in the atmosphere, such as those related to climate change or the effects of human activity on the environment.

Remote sensing can be passive and active. Passive remote sensing involves measuring the radiation emitted or reflected by the Earth's atmosphere and surface in various spectral regions, such as visible, infrared, and microwave. Examples of passive remote sensing instruments include the Moderate Resolution Imaging Spectroradiometer (MODIS) on NASA's Terra and Aqua satellites and the Atmospheric Infrared Sounder (AIRS) on NASA's

4 https://earthobservatory.nasa.gov/images/86257/an-epic-new-view-of-earth

5 Lagrangian Point 1 is one of five special points in a two-body system (such as the Earth and the Sun) where the gravitational forces of the two large bodies, along with the centrifugal force due to the orbit of a smaller object, balance out. This balance allows a smaller object to remain in a stable or quasi-stable position relative to the two larger bodies. L1 is located on the line connecting the two large bodies (e.g. the Earth and the Sun) and lies between them. For the Earth–Sun system, L1 is about 1.5 million kilometers from Earth, or about 1/100th the distance between the Earth and the Sun.

Aqua satellite. A schematic illustrating passive remote sensing is shown in Figure 1.6. Active remote sensing involves transmitting electromagnetic energy into the atmosphere and measuring the backscattered signals. CALIPSO (Cloud-Aerosol Lidar and Infrared Pathfinder Satellite Observation)[6] uses lidar technology to analyze the vertical structure of clouds and aerosols, providing essential data for studying their properties and distribution. CloudSat[7] employs radar to penetrate cloud layers and yield detailed information about cloud structure and composition. The scheme of passive remote sensing is depicted in Figure 1.7.

The acquired remote sensing data can then be used to retrieve a variety of information related to the Earth's atmosphere, including temperature profiles, air pressure profiles,

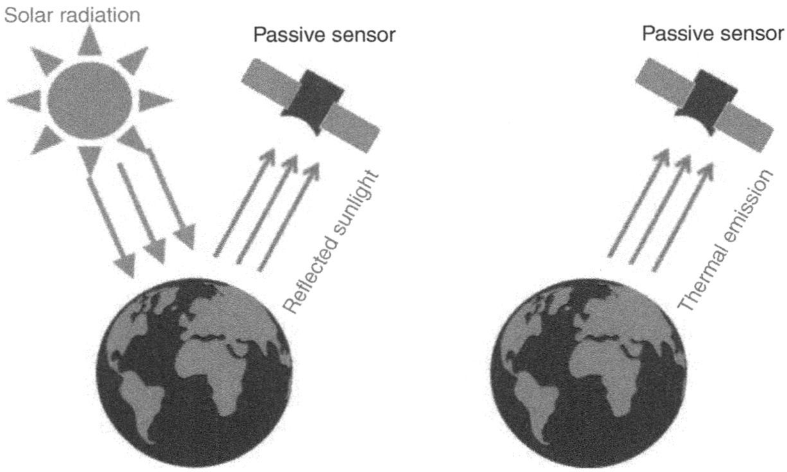

Figure 1.6 Schematic illustrating passive remote sensing: (left) a sensor onboard a satellite measuring reflected sunlight; (right) a space-based sensor measuring terrestrial thermal emission.

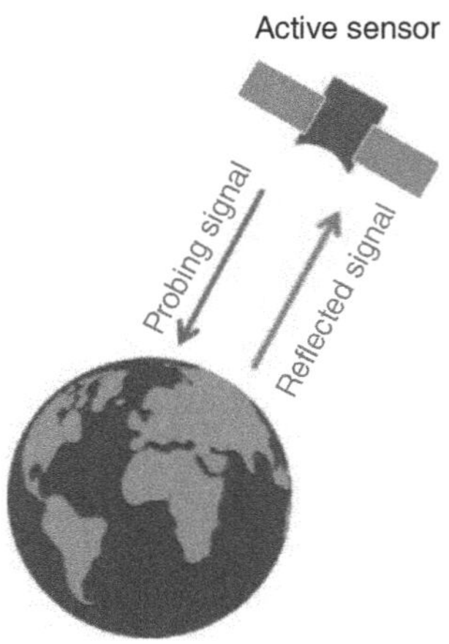

Figure 1.7 Schematic illustrating active remote sensing.

6 https://www-calipso.larc.nasa.gov/
7 https://cloudsat.atmos.colostate.edu/

water vapor content, cloud properties, aerosol characteristics, and trace gas concentrations. This information is essential for understanding the physical and chemical processes that occur in the atmosphere and how they impact climate, weather, and air quality. Additionally, remote sensing data can be used for monitoring natural disasters, such as wildfires and volcanic eruptions, and for assessing the impact of human activities on the environment, such as air pollution from industrial emissions and deforestation.

In remote sensing, the quantity of interest is not directly measured by the instrument. The received signal undergoes several stages of processing before the necessary information about a certain parameter can be extracted and used in various applications. To describe these processing stages, a standardized set of "levels" has been adopted. Each level represents a sub-process with well-defined inputs and outputs that can be considered independently. The levels are as follows:

Level 0: This is the raw, unprocessed data collected by the sensor, usually in the form of counts or volts.

Level 1: At this stage, the Level 0 data is converted into energy units (typically radiance).

Level 2: The retrieved geophysical parameters, such as aerosol/cloud parameters and trace gas concentrations, are obtained at the same resolution and location as the Level 1 data.

Level 3: The Level 2 data is mapped onto uniform space/time grid scales, allowing data from different sources to be combined.

Level 4: The final stage involves using the Level 3 data to derive conclusions, results, and other derived data.

1.6 Radiative Transfer Models in Atmospheric Remote Sensing

Forward and inverse problems represent fundamental concepts in mathematical modeling and computational sciences, including areas like physics, engineering, remote sensing, and medical imaging. Understanding the distinction between these two types of problems is crucial for developing models and algorithms to interpret data and infer properties of physical systems (see Figure 1.8).

Generally speaking, forward problems start with the causes (the parameters and conditions) and compute the effects (the observables). The forward problem is generally well-posed; given the input parameters, the outcome is predictable and unique. This process is foundational in simulation studies where the aim is to understand how a system behaves under specific conditions. A forward model simulates observables, such as measured spectra, for a given set of model parameters. This approach allows researchers and scientists to

Figure 1.8 Forward and inverse problems.

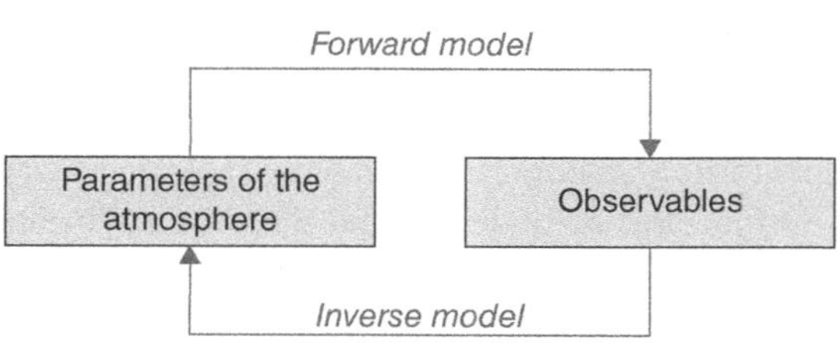

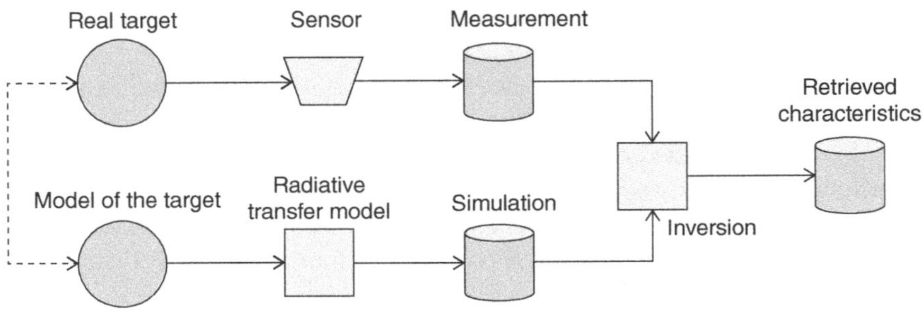

Figure 1.9 Radiative transfer model in remote sensing.

predict the outcome of a system under specific conditions, effectively turning theoretical model parameters into directly comparable real-world measurements. In remote sensing, forward modeling involves computing the radiances for a given atmospheric state, typically by converting physical parameters of the atmosphere to optical parameters and then using a radiative transfer model to compute the scattered radiance field (Ustinov, 2014). The forward modeling process is used to generate simulated data that can be compared (fitted) to actual measurements to infer properties of the underlying system. The key component of the forward model in remote sensing is a radiative transfer model (RTM) (see Figure 1.9). RTMs are used to calculate the scattered radiance field and thereby estimate the signal received by a sensor.

Inverse problems, on the other hand, seek to determine the unknown parameters or characteristics of a system from observed outcomes or data (Rodgers, 2000; Doicu et al., 2010). Inverse problems are often much more challenging than forward problems because they involve inferring causes from effects. Inversion can be a challenging problem, as it often requires the use of specialized algorithms to solve the inverse problem, and may require additional information or assumptions to be made about the system being studied. This process can be ill-posed, meaning that the solution might not be unique or may not depend continuously on the data; small changes in the observed data can lead to large changes in the inferred parameters. In atmospheric remote sensing, the extraction of information about atmospheric constituents and surface properties is referred to as retrieval.

RTMs can be used to infer the concentrations of various atmospheric constituents such as water vapor, carbon dioxide, ozone, and various atmospheric pollutants. The estimation of aerosol optical properties, including aerosol optical depth, size distribution, and single-scattering albedo, is crucial for understanding aerosol effects on climate, air quality, and visibility. RTMs can be used to retrieve cloud optical thickness, cloud top height, cloud particle size distribution, and cloud phase (liquid, ice) from remote sensing observations. Cloud properties are important for understanding cloud radiative effects and their influence on climate. RTMs can infer surface properties such as surface reflectance, emissivity, and temperature. This information is essential for various applications, including land surface monitoring, climate modeling, and agriculture. To obtain surface imagery without atmospheric effects, it is necessary to perform atmospheric correction (Richter, 1990) by removing the effects of the atmosphere from the signal obtained through remote sensing.

1.7 Electromagnetic Spectrum

Taking the curl of Eq. (1.7), and substituting Eq. (1.8) into it, along with the application of the vector identity

$$\nabla \times (\nabla \times \mathbf{E}) = \nabla(\nabla \cdot \mathbf{E}) - \nabla^2 \mathbf{E}, \tag{1.13}$$

leads us to derive the wave equation for the electric field:

$$\nabla^2 \mathbf{E} - \epsilon_0 \mu_0 \frac{\partial^2 \mathbf{E}}{\partial t^2} = 0. \tag{1.14}$$

A typical solution to Eq. (1.14) is a plane wave expressed as:

$$\mathbf{E}(\mathbf{r}, t) = \mathbf{E_0} \cos(2\pi \nu t - \mathbf{k} \cdot \mathbf{r}), \tag{1.15}$$

where $\mathbf{E}_0$ is the amplitude, $\mathbf{r}$ is the position vector in space, t is the time, ν is the frequency, and $\mathbf{k}$ is the wave vector, indicating the direction and wavelength of the wave.

In remote sensing applications, these principles of wave propagation are utilized extensively. Various parts of the electromagnetic spectrum are used to gather data about the Earth's surface and atmosphere. Each segment of the spectrum offers distinct characteristics and is suitable for different applications, as detailed in Table 1.3. At one end of the spectrum are the high energy, short wavelength gamma rays, and X-rays, which are typically not used in terrestrial remote sensing due to atmospheric opacity. These high-energy wavelengths can be useful for studying crystal structures and surface states in laboratory measurements, or for studying planetary surfaces with no atmosphere, such as the Moon. X-rays are used for medical diagnostics. One of the primary reasons for the widespread use

Table 1.3 Electromagnetic spectrum regions.

Name	Wavelength	Frequency	Photon energy
Gamma ray	0.01 nm	30 EHz	124 keV
X-ray	0.01–10 nm	30 PHz–30 EHz	1.24 eV–124 keV
Extreme ultraviolet	10–120 nm	2.5–30 PHz	10.35–124 eV
Far ultraviolet	120–200 nm	1.5–2.5 PHz	6.21–10.35 eV
Middle ultraviolet	200–300 nm	1.0–1.5 PHz	4.14–6.21 eV
Near ultraviolet	300–400 nm	750 THz–1.0 PHz	3.1–4.14 eV
Visible	400–700 nm	420–750 THz	1.77–3.1 eV
Near-infrared	700–3500 nm	86–420 THz	0.35–1.77 eV
Thermal infrared	3.5–100 μm	3–86 THz	0.012–0.35 eV
Microwaves	1 mm–1 m	300 MHz–300 GHz	1.24 μeV–1.24 meV
Radio waves	1 mm–10000 km	30 Hz–300 GHz	1.24 meV–124 feV
Extremely low frequency	10000 km	30 Hz	12.4 feV

of X-rays in medical imaging is their ability to penetrate the body without scattering, thus producing highly detailed and focused images. While X-rays can be absorbed by certain materials, their ability to travel through soft tissue and bone allows them to provide a clear and sharp view of the internal structures of the body.

Going further along the spectrum, we find the ultraviolet (UV) region, which can be divided into three subregions: extreme UV, far UV, and near UV. Extreme UV is generated by the solar corona and is highly absorbed in air, ionizing O_2 and N_2, while far UV is absorbed by O_2 and middle UV is mostly absorbed by the ozone layer, which is located around 25 km above sea level. The near UV can reach Earth's surface and can have harmful effects on human health, causing skin damage, cataracts, immune system suppression, and genetic damage, resulting in skin cancer. However, moderate sun exposure is important for human health, as it stimulates the synthesis of vitamin D in the skin.

The visible range (400–700 nm) is the part of the electromagnetic spectrum that humans can see. It is divided into seven subregions: violet, indigo, blue, green, yellow, orange, and red. This region is used extensively in imaging techniques, mapping, and multispectral photography. Clouds and aerosols have a strong impact on the spectrum in this region, and this feature can be used for both passive and active remote sensing techniques.

The near-infrared (NIR) region, which spans from approximately 700 to 1400 nm, contains strong molecular absorption bands of water vapor, oxygen, and carbon dioxide. The solar light reflectance in the gaseous absorption bands can be used for various applications, including the estimation of cloud/aerosol layer altitudes. The thermal infrared (TIR) region is used for thermal imaging and remote sensing of land and surface water temperatures. The emission of thermal radiation is an important mechanism of radiation cooling, and so, this part of the spectrum has a strong impact on the global climate.

At the other end of the spectrum, we find the microwave region, which spans from approximately 1 mm to 1 m in wavelength. Microwave radiation can penetrate through clouds, making it useful for remote sensing applications in all weather conditions and also at night. Passive microwave remote sensing detects the naturally emitted microwave energy, which is small compared to the optical region. The field of view of the instrument should be large enough so that the spatial resolution of the microwave sensors is lower than those operated in the visible region. Active microwave sensors can be imaging and non-imaging, with imaging sensors called RADAR (radio detection and ranging), which can be used for imaging at any time, day or night, and in most weather conditions. The time between the transmitted and reflected signals, and the corresponding phase difference, determines the distance to the target. Non-imaging sensors, such as the Meteorological Temperature Profiler MTP-5, serve the same purpose.

In our book, we consider radiative transfer models that are typically used in the UV, VIS, NIR, and SWIR regions of the electromagnetic spectrum. These spectral regions are used extensively in remote sensing applications, and understanding the unique characteristics and applications of each region is crucial for developing effective remote sensing methods. The spectra at the top of the atmosphere and at the sea level are different due to the attenuation of radiation as it passes through the atmosphere, as shown in Figure 1.10. The top of the atmosphere spectrum is the amount of energy per unit area per unit of time that is incident on the outermost layer of the Earth's atmosphere. This spectrum includes all types of radiation, such as ultraviolet, visible, and infrared radiation, as well as X-rays and gamma

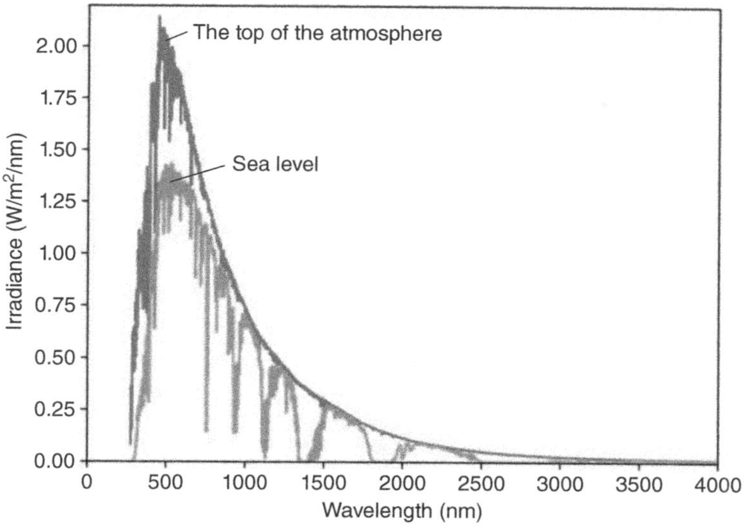

Figure 1.10 Examples of solar spectra at the top of the atmosphere and at sea level.

rays. As the radiation travels through the atmosphere, it encounters various molecules such as oxygen, water vapor, ozone, carbon dioxide, and aerosols, which can scatter and absorb the radiation. As a result, the spectrum of radiation at sea level is different from the spectrum at the top of the atmosphere. The amount of absorption and scattering varies with wavelength, so different regions of the electromagnetic spectrum are affected differently. For example, ultraviolet radiation is mostly absorbed by ozone in the upper atmosphere, so very little ultraviolet radiation reaches the Earth's surface. In contrast, visible light is not significantly absorbed or scattered, so it can penetrate the atmosphere and reach the surface relatively unimpeded. The essential portion of infrared radiation, on the other hand, is absorbed by water vapor and carbon dioxide, which contributes to the greenhouse effect and can lead to warming of the lower atmosphere. Radiative transfer models can help to predict how the spectra will be modified as radiation passes through different layers of the atmosphere, taking into account factors such as the composition and density of atmospheric gases, as well as the presence of aerosols and clouds.

1.8 Why Do We Need Analytical Models in Radiative Transfer?

Recent developments in optics, sensor design, and measurement techniques have greatly enhanced the features of remote sensing sensors, in particular, the spatial resolution of the measurements. The spatial resolution of remote sensing images refers to the smallest object that can be detected on the Earth's surface. Advances in optics, sensor technologies, and measurement techniques have led to a significant reduction in the size of the smallest detectable feature. This improvement in spatial resolution means that finer-scale changes can be monitored over time, enhancing applications in urban planning, agriculture, forestry, disaster management, and environmental monitoring.

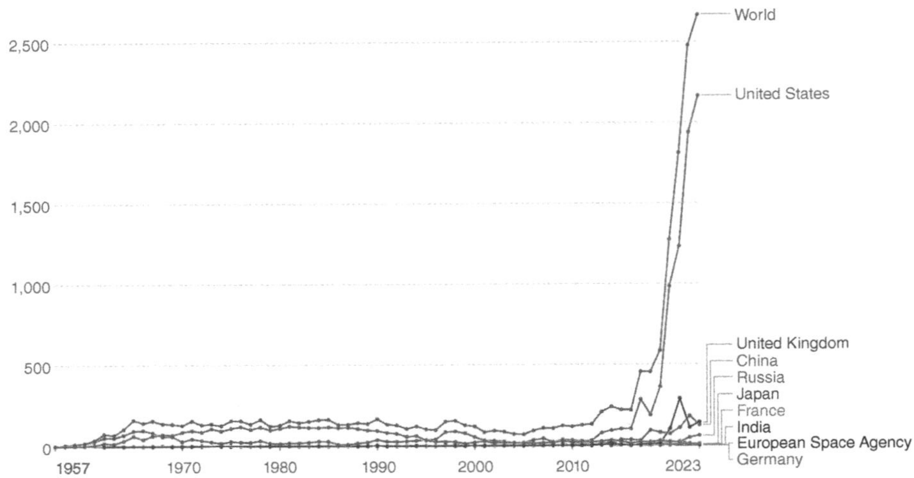

Figure 1.11 Annual number of objects launched into space. *Source*: Data taken from Mathieu and Roser (2022) (licensed under CC BY 4.0).

There is a significant change taking place in the industry regarding the number of satellites and their market value. There is a forecasted increase of 4.5 in satellite demand over the next 10 years, which means an average of about 1800 satellites will be launched each year, compared to about 400 over the last decade.[8] According to the Union of Concerned Scientists (UCS) Satellite Database,[9] the number of satellite launches increases exponentially, as shown in Figure 1.11.

The high volume of data generated by remote sensing can be overwhelming, making it challenging to analyze and interpret. Traditional approaches are often inadequate to deal with large amounts of data, as the amount of satellite data is growing faster than the computational power (Ma et al., 2015). The performance bottleneck in the processing chain is radiative transfer calculations that are computationally expensive and time-consuming. The remote sensing data is recognized as Big Data (Liu, 2015) since it satisfies Doug Laney's 3V criterion (Patgiri and Ahmed, 2016): significant growth in volume, velocity, and variety. Volume refers to the physical amount of data, velocity to the speed at which the data is created, and variety to the different forms of data. New computational paradigms are required to cope with these high-efficiency requirements, including integrating data from different sources in various locations, distributing data across massively parallel platforms for processing, and providing fast access to remote sensing data to users worldwide.

To address this challenge, analytical models can provide useful information while reducing the computational burden. Approximate models are simplified versions of the original

8 https://digital-platform.euroconsult-ec.com/product/satellites-to-be-built-launched-new/
9 https://www.ucsusa.org/resources/satellite-database

models that capture the essential features of the physical system. They are designed to provide a reasonable approximation of the physical model while reducing the complexity and computational requirements. The so-called surrogate models can be applied (Brence et al., 2022). Essentially, a surrogate model is a computationally efficient approximation of a complex physical model or a function. This method is particularly beneficial for critical applications, as it accelerates the simulation process, freeing up more computational time to explore a larger parameter space or improve the accuracy of methods through more iterations. Approximate radiative transfer models can provide a relatively fast and accurate approximation of the radiation field, which can then be used to derive insights and understanding of the problem (Doicu et al., 2022). Moreover, they can provide a solution in an analytical form, making it easier to analyze and understand the behavior of the radiation field under different conditions, such as varying atmospheric parameters or surface properties. This can be particularly important in fields such as climate science or astrophysics, where small changes in atmospheric or environmental conditions can have significant effects on the radiation field. Some approximate models can significantly improve the accuracy of the dimensionality reduction and machine learning models, as discussed further in this book.

1.9 Radiative Transfer and Climate Modeling

Radiative transfer plays a fundamental role in climate modeling by simulating the transfer of electromagnetic radiation through the Earth's atmosphere and surface. Climate models utilize RTMs to understand how incoming solar radiation, outgoing terrestrial radiation, and energy exchanges within the atmosphere influence Earth's climate system. The Earth's surface absorbs solar radiation and re-emits it as longwave (infrared) radiation. Radiative transfer models calculate the absorption and emission of terrestrial radiation by greenhouse gases such as water vapor, carbon dioxide, methane, and ozone. These models also consider the scattering and absorption of infrared radiation by clouds and aerosols. Understanding the balance between incoming solar radiation and outgoing terrestrial radiation is crucial for determining Earth's energy budget and establishing the global temperature distribution. Radiative transfer models simulate the greenhouse effect, which is the process by which certain atmospheric gases trap heat and warm the Earth's surface. These models quantify the radiative forcing exerted by greenhouse gases, which is the imbalance between incoming solar radiation and outgoing terrestrial radiation caused by the presence of these gases. By incorporating information on greenhouse gas concentrations, radiative properties, and atmospheric dynamics, climate models can assess the magnitude of the greenhouse effect and its impact on global temperature changes.

Radiative transfer models also account for feedback mechanisms that can amplify or dampen climate change. Climate models simulate these feedback processes to project future climate scenarios under different greenhouse gas emission scenarios and assess the uncertainties associated with climate projections. Two distinct types of feedback loops can be identified: positive and negative. Positive feedback amplifies changes, driving systems away from equilibrium and often leading to exponential growth or instability. In contrast, negative feedback mitigates changes, promoting balance and stability by counteracting deviations and steering systems back toward equilibrium. Positive feedback mechanisms within the atmosphere are those that magnify initial alterations in the climate system, thereby intensifying either warming or cooling trends. Conversely, negative feedback mechanisms serve to mitigate climate perturbations, often restoring the climate system to its original state by counteracting the initial changes. The concept of positive and negative feedback loops in atmospheric science is closely linked to the work of early climate scientists and systems theorists. One of the most notable figures in formalizing these ideas within the context of climate dynamics is Hansen et al. (1981). His work on the effects of increasing greenhouse gases emphasized the role of feedback mechanisms in climate systems, especially how positive feedback loops could amplify global warming. Several feedback mechanisms are important for climate modeling, especially in the context of analysis of the robustness of the climate system, including:

Water vapor feedback: The atmosphere warms due to increased greenhouse gas concentrations. This leads to higher evaporation rates from the oceans and other water bodies. Since water vapor is itself a greenhouse gas, the increased atmospheric water vapor content further enhances the greenhouse effect, causing additional warming. This feedback loop amplifies the initial temperature increase, contributing to enhanced global warming (Hall and Manabe, 1999).

Cloud feedback: The atmosphere warms due to increased greenhouse gas concentrations. This leads to higher evaporation rates from the oceans and other water bodies. Since water vapor is itself a greenhouse gas, the increased atmospheric water vapor content further enhances the greenhouse effect, causing additional warming. This feedback loop amplifies the initial temperature increase, contributing to enhanced global warming (Hall and Manabe, 1999).

Ice-albedo feedback: The reflectivity of the Earth's surface, known as albedo, plays a crucial role in regulating the planet's temperature. Ice and snow have high albedo, meaning they reflect a significant portion of incoming solar radiation back to space. However, as temperatures rise, ice and snow cover decrease, exposing darker surfaces such as ocean or land. This reduces the overall albedo, leading to increased absorption of solar radiation and further warming. The loss of Arctic sea ice and the melting of glaciers are examples of ice-albedo feedback mechanisms that contribute to amplified warming in polar regions (Budyko, 1969). The darkening of ice and snow surfaces due to different pollutants, including dust and black carbon, contributes to the global albedo reduction as well (Hansen and Nazarenko, 2003; Flanner et al., 2007).

Permafrost feedback: Permafrost, frozen soil found in Arctic and subarctic regions, contains vast amounts of organic matter that can decompose and release greenhouse gases like methane and carbon dioxide when thawed. As temperatures rise, permafrost thaws, accelerating the decomposition of organic material and releasing additional greenhouse gases into the atmosphere. This process further enhances the greenhouse effect, leading to more warming and additional permafrost thawing (Schuur et al., 2015) – a positive feedback loop.

Vegetation feedback: Changes in temperature and precipitation patterns can affect vegetation growth, distribution, and productivity. For example, warmer temperatures may increase the rate of photosynthesis and plant growth in some regions, leading to greater carbon uptake by vegetation. However, in other cases, warming and changes in precipitation may lead to drought stress, wildfires, and dieback of forests,[10] reducing carbon sequestration capacity. These changes in vegetation cover and carbon cycling can influence atmospheric carbon dioxide levels, contributing to positive feedback in the climate system (Cheng et al., 2021).

Water vapor and lapse rate feedback: While increased atmospheric temperatures can lead to higher water vapor content due to enhanced evaporation, this also leads to greater cloud formation. Clouds have both warming (by trapping outgoing longwave radiation) and cooling (by reflecting incoming solar radiation) effects. Additionally, as the atmosphere warms, the lapse rate (the rate of temperature decreases with altitude) may change, which can influence cloud formation and distribution. Changes in cloud cover and the lapse rate can act as negative feedback, ultimately counteracting some of the initial temperature increases.

Cloud albedo effect: Changes in cloud properties, such as cloud cover, altitude, and particle size, can influence the amount of solar radiation reflected back to space. Low-level clouds, such as stratocumulus, have a cooling effect because they reflect a significant portion of incoming solar radiation. As temperatures rise, increased cloud cover or changes in cloud properties may enhance this cooling effect, offsetting some of the initial warming (Mueller et al., 2011).

1.10 Remote Sensing of Trace Gases

Remote sensing of trace gases is a critical area of study in atmospheric science, environmental monitoring, and climate research. It involves the detection and quantification of gases present in the atmosphere in small quantities, which, despite their low concentrations, can have significant impacts on the environment and human health. The concentrations of different gases in the Earth's atmosphere are listed in Table 1.4.

Despite the relatively low concentrations of trace gases in the atmosphere, they can be effectively detected by analyzing the spectrum of electromagnetic radiation at specific wavelengths where the absorption by these gases is particularly strong. Each gas is characterized by a unique set of absorption lines, often referred to as its spectral "fingerprints." These unique patterns are critical in identifying and quantifying different gases, as they allow for precise discrimination between various atmospheric components.

To illustrate this, let's conduct a thought experiment. Consider the passage of light through the atmosphere, but imagine that the atmosphere consists solely of one gas. Then, depending on the wavelength, a different fraction of radiation will pass through for each gas. This fraction is called the transmittance. The corresponding transmittance spectra for four gases are shown in Figure 1.12. Regions with transmittance close to 1 are called atmospheric windows because the atmosphere is transparent at these wavelengths, allowing radiation to pass through with minimal absorption. Conversely, regions where the transmittance is significantly lower indicate wavelengths that are strongly absorbed by the atmospheric gases.

10 The phenomenon of a stand of trees losing health and dying without an obvious cause.

Table 1.4 Gaseous concentrations in the Earth's atmosphere.

Gas	Concentration	Comments
Nitrogen (N_2)	78.08%	Most abundant atmospheric gas, inert
Oxygen (O_2)	20.95%	Essential for respiration and combustion
Argon (Ar)	0.93%	Inert
Carbon dioxide (CO_2)	420 ppm	Greenhouse gas, important for photosynthesis
Water vapor (H_2O)	Variable	Critical for weather systems, greenhouse gas
Neon (Ne)	18 ppm	Inert
Helium (He)	5.2 ppm	Light, inert
Methane (CH_4)	1.8 ppm	Potent greenhouse gas affects climate change
Krypton (Kr)	1 ppm	Inert
Hydrogen (H_2)	0.5 ppm	Lightest gas
Xenon (Xe)	0.09 ppm	Inert
Ozone (O_3)	0.1–10 ppb	Absorbs UV radiation, pollutants at ground level

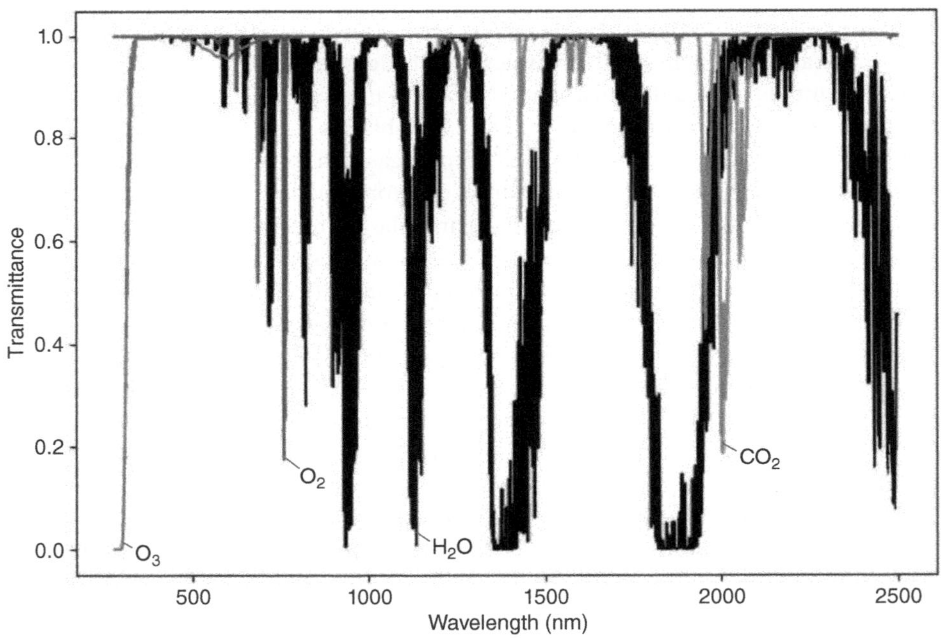

Figure 1.12 Transmittances for different gases in the spectral range 400–2500 nm.

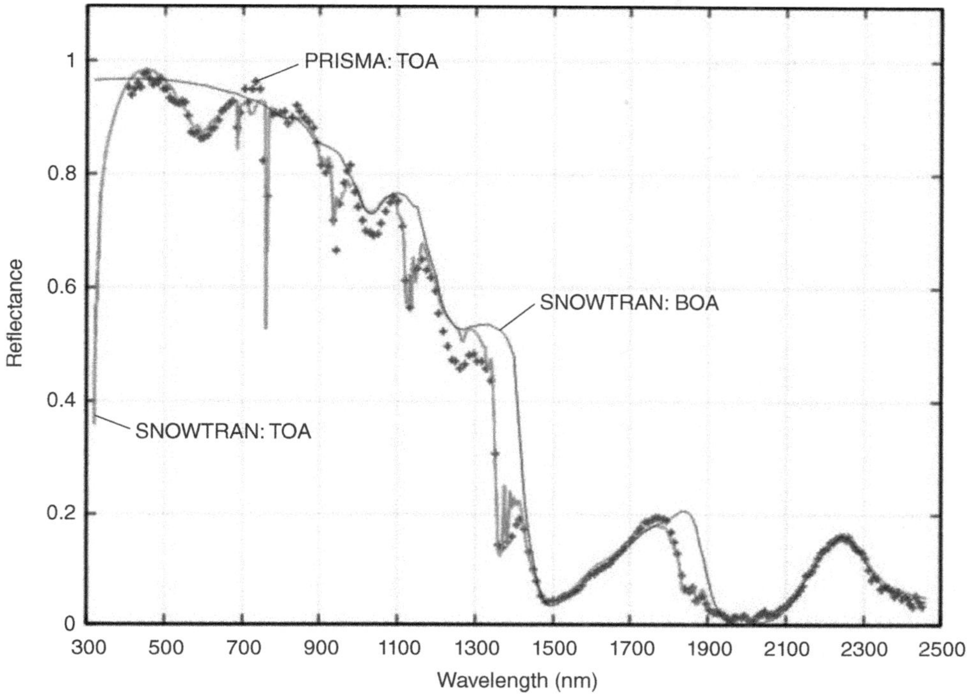

Figure 1.13 Measured PRISMA top-of-atmosphere (TOA) spectral reflectance (crosses) for a solar zenith angle of 57.6° and nadir observation over the underlying snow surface in Antarctica. The lines show results from theoretical modelling of TOA and bottom of atmosphere (BOA) reflectances with a spectral resolution of 1 nm using SNOWTRAN software. *Source:* Adapted from Kokhanovsky et al. (2024) (licensed under CC BY 4.0).

Measurements taken at characteristic wavelengths are extremely sensitive to the presence of specific gaseous species. This sensitivity enables the detection and analysis of trace gases even at minute concentrations, making it an invaluable tool in environmental monitoring, atmospheric chemistry, and climate science. The technique not only helps in tracking pollution and studying atmospheric conditions but also plays a crucial role in monitoring changes related to global warming and other environmental issues.

We show the top-of-atmosphere reflectance measured by PRISMA hyperspectral spaceborne instrument[11] on December 21, 2022 (00:14UTC) at the point with coordinates (76.14N, 129.93E) over underlying snow surface in Figure 1.13. The absorption of reflected light by various molecules (O_2, O_3, H_2O) is clearly identified. The depth of absorption lines as seen in the reflected light can be used to retrieve the total columns of respective gases.

11 https://www.eoportal.org/satellite-missions/prisma-hyperspectral

1.11 Remote Sensing of Clouds

Clouds, composed of microscopic water droplets and/or ice crystals, play a crucial role in Earth's hydrological cycle and climate system. They significantly influence radiation processes, atmospheric dynamics, chemistry, and thermodynamics. Clouds play a crucial role in weather prediction and climate models. Understanding their types and behaviors helps meteorologists predict weather changes and climatologists understand the Earth's climate system.

Clouds are classified into various types based on their appearance, altitude, and the physical processes that lead to their formation. Table 1.5 provides an overview of common cloud types alongside typical models used to study them. Clouds can be classified based on their altitude, which also corresponds to differences in cloud top pressure and thickness. These classifications include:

1. **High-level clouds** (5–13 km): Typically found at altitudes of 5–13 km, these clouds include types such as cirrocumulus, cirrus, and cirrostratus, which are composed mainly of ice particles.
2. **Mid-level clouds** (2–7 km): Existing between 2 and 7 km, these clouds include altocumulus, altostratus, and nimbostratus, which may be composed of water droplets, ice particles, or a mixture of both.
3. **Low-level clouds** (0–2 km): Found at altitudes up to 2 km, these include stratus, cumulus, cumulonimbus, and stratocumulus, which are generally water cloud types.

It is important to note that these classifications are not universal, and different meteorological sources may use alternative schemes based on regional climatic conditions and research needs.

The optical properties of clouds are primarily determined by the characteristics of the particles they are composed of. This includes the particle size, shape, and the complex refractive index, which affect how light interacts with cloud particles.

Table 1.5 Common cloud types and corresponding models.

Cloud type	Description	Model used
Cumulus	Fluffy, heap-like clouds with flat bases	Cumulus parameterization models
Stratus	Layered, uniform clouds covering the sky	Stratiform cloud models
Cirrus	Thin, wispy clouds at high altitudes	High-altitude cloud models
Nimbus	Rain-bearing clouds that are thick and dark	Precipitation models
Cumulonimbus	Tall, dense clouds with significant vertical growth	Severe weather models

Water clouds are often modeled as collections of spherical water droplets, adhering to specific size distributions. One common model used is the modified gamma distribution, represented mathematically as follows (Deirmendjian, 1969):

$$f(r) = ar^{\alpha}e^{-br^{\gamma}}, \tag{1.16}$$

where a, α, b, and γ are parameters that define the shape of the size distribution. This distribution helps in describing the variability in droplet sizes within a cloud.

Cirrus clouds contain nonspherical ice particles. This introduces significant computational challenges in simulating how radiation interacts with these clouds. Nonspherical particles require more complex models to accurately represent their interaction with light, impacting the accuracy of remote sensing data and climate models (Yang and Liou, 2016). Retrieving cloud properties using the molecular oxygen absorption bands is a critical technique in remote sensing, especially in atmospheric and environmental sciences. This approach utilizes the strong absorption features of molecular oxygen in the optical range to assess cloud characteristics such as cloud top pressure, cloud cover, and optical thickness. Several Earth-observing satellites and instruments are equipped to analyze the O_2 A-band for the retrieval of cloud properties, such as MODIS, GOME-2, and OMI. These instruments provide global data that are crucial for daily weather forecasting, climate modeling, and atmospheric research.

Retrieving cloud properties using the O_2 A-band typically involves analyzing the changes in the depth and shape of the absorption band as light passes through the atmosphere and is reflected back by clouds. This technique is based on the principle that the amount of light absorbed by oxygen varies with the path length:

- **Spectral analysis:** Instruments measure the spectral characteristics of sunlight that has traveled through the atmosphere and reflected off the cloud tops. The absorption features in the O_2 A-band are analyzed to determine the path length of light.
- **Cloud top pressure:** The amount of absorption in the O_2 A-band can be used to estimate cloud top pressure. Light interacting with higher clouds (at lower pressures) will show different absorption characteristics compared to light interacting with lower clouds (at higher pressures).
- **Optical thickness and cloud cover:** By analyzing the intensity and width of the O_2 A-band, researchers can infer the cloud optical thickness and overall cloud cover. Thicker clouds and denser cloud cover affect the absorption features differently compared to thin or sparse clouds.

Figure 1.14 illustrates the variations in radiance at the top of the atmosphere within the O_2 A-band in response to changes in cloud optical thickness and cloud top height, with a consistent cloud geometrical thickness set at 1 km. The wings of the band are predominantly sensitive to variations in cloud optical thickness, whereas the central portion of the band is more responsive to changes in the cloud's vertical position.

Radiative transfer simulations that incorporate scenarios with clouds can benefit from considering the medium as optically thick. In such cases, the cloud acts as a barrier that obscures the underlying surface, along with the spectral signatures of trace gases located beneath it. Often, clouds can be approximated as diffuse reflective surfaces, simplifying the computational models used to analyze and predict their impact on observed radiation from above.

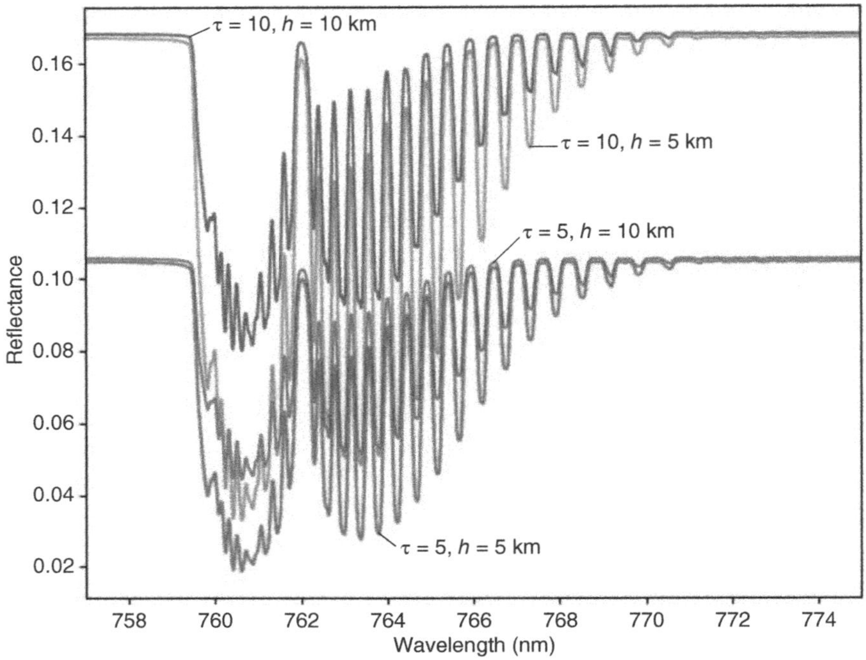

Figure 1.14 TOA radiance in the O_2 A-band for different values of cloud optical thickness τ and cloud top height h. The spectra are convolved with a slit function that has a full width at half maximum of 1 nm. The viewing zenith angle is 40°, and the solar zenith angle is 0°. The computations are performed using the PYDOME radiative transfer model (Efremenko et al., 2023).

1.12 Remote Sensing of Atmospheric Aerosol

Atmospheric aerosol is composed of liquid and solid microscopic particles with average radii in the range of 0.1–3 μm. These tiny particles are responsible for cloud formation and have important climatic impacts. The aerosol optical thickness is usually small (in the range 0.1–0.5) in the visible spectrum, with smaller values in the polar regions and extremely large values during heavy forest fires, volcanic eruptions, and dust storms. The single-scattering albedo is close to 1 in the visible range in the majority of cases. The parameters of scientific interest are particle size distributions (e.g. in fine and coarse modes) and also complex spectral refractive index of particles, the columnar concentration of particles, the position of aerosol layer in atmosphere, fine particulate matter at the ground ($PM_{2.5}$), spectral aerosol optical thickness, and single-scattering albedo. We show the global spatial variation of the parameter $PM_{2.5}$ derived from satellite top-of-atmosphere measurements before and during COVID-19 pandemic in Figure 1.15. The decrease in the value of fine particulate matter with diameters below 2.5 μm at the ground surface is clearly seen. It is related to the decline in industrial production during COVID-19 pandemic. This also means that cloud nuclei concentrations have been reduced, which have reduced cloud cover and resulted in heating of the Earth system with important climatic impacts due to both atmospheric aerosol and clouds.

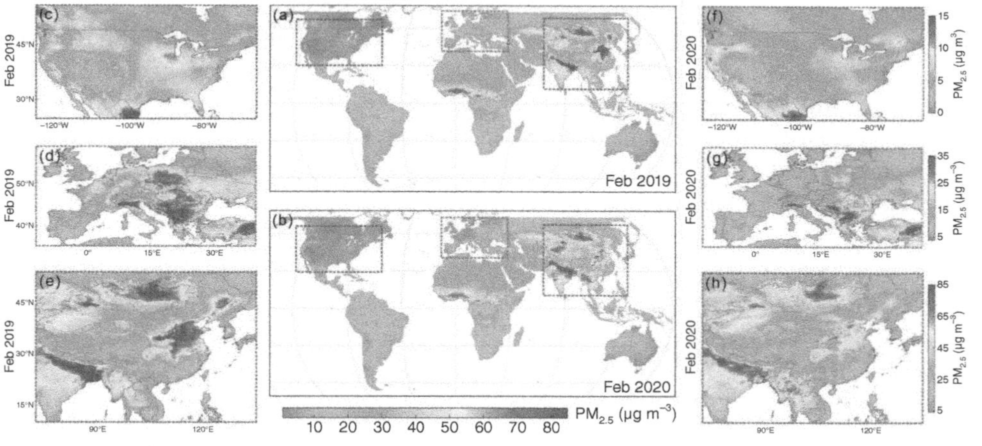

Figure 1.15 Global distribution of fine particulate matter at the surface before and during the COVID-19 pandemic. *Source*: Adapted from Bai et al. (2024) (licensed under CC BY 4.0).

Table 1.6 Types of aerosols, their sources, and radiative effects.

Type	Source	Radiative effect
Oceanic aerosol	Sea spray from the ocean surface, salt particles, and organic matter	Primarily negative forcing: reflects sunlight
Dust aerosol	Wind-blown dust from deserts, arid regions, and land surface disturbances	Mixed forcing: reflects and absorbs sunlight
Volcanic aerosol	Eruptions release sulfur dioxide (SO_2), forming sulfate aerosols	Strong negative forcing: reflects sunlight
Smoke aerosol	Biomass burning, forest fires, and human activities like agricultural burning	Mixed forcing: absorbs and reflects sunlight
Biogenic aerosols	Emissions from plants, trees, and other biological processes	Negative forcing: reflects sunlight
Background water-soluble aerosol	Naturally occurring water droplets and dissolved compounds in the atmosphere	Negative forcing: reflects sunlight

Maps, as shown in Figure 1.15, can be derived from the analysis of reflected solar light intensity and polarization in different regions of the planet with the use of radiative transfer models to be discussed in this work. This underlines the importance of the topic considered in this book.

The atmospheric aerosol is usually classified in terms of its origin. Their radiative effect is also different. Some types of aerosols together with their effects are summarized in Table 1.6.

Usually it is assumed that there are two fractions of atmospheric aerosol: fine mode and coarse mode aerosol fractions, which are characterized by the lognormal volume particle size distribution with the median radius in the range 0.12–0.15 μm for the fine mode and 1.9–3.3 μm for the coarse mode (Dubovik et al., 2002). The best retrievals of aerosol properties are achieved if multi-angular spectropolarimetry is used (Dubovik et al., 2011; Kokhanovsky et al., 2015). The aerosol top height position can be derived from lidar measurements or oxygen A-band spectrometry (Jänicke et al., 2023; Huang et al., 2015).

2

Radiative Transfer Equation

2.1 Introduction to Radiative Transfer Theory

Radiative transfer theory (RTT) deals with energy transfer in the form of electromagnetic radiation. Early attempts to describe the physical phenomenon quantitatively resulted in the geometrical optics laws, the extinction law of Bouguer, and the attenuation of light intensity with the inverse square of distance. However, these laws cannot explain the transfer of the radiation through a turbid medium, e.g. the angular distribution of transmitted light intensity, which tends to become more isotropic as the light passes through the medium. The first studies of the radiative transfer equation (RTE) have been performed by Lommel (1887) and Chwolson (1889). RTT has found numerous applications in many branches of modern science and technology, including atmospheric physics and astrophysics (Chandrasekhar, 1950; Sobolev, 1956, 1975; van de Hulst, 1980; Liou, 2002a; Mishchenko et al., 2007). In particular, in atmospheric remote sensing, RTT is used to compute the amount of solar radiation reflected by the Earth-atmosphere-underlying surface system or transmitted through its atmosphere. This information is used to retrieve information about the atmospheric composition of our planet and to provide a deeper understanding of the Earth's climate system. In astrophysics, the light coming from distant stellar objects is analyzed to understand their physical and chemical properties. It is no wonder that radiative transfer was the focus of outstanding astrophysicists, such as Ambartsumyan (1942), Chandrasekhar (1950), Sobolev (1975), and others. A comprehensive review of discussions about the foundations of radiative transfer can be found elsewhere (Mishchenko, 2008, 2013; Dolin, 1964; Apresyan and Kravtsov, 1996; Kravtsov and Apresyan, 1996; Aizenberg and Budak, 2018).

Radiance (specific intensity) I is a fundamental quantity in radiometric physics. It is usually denoted as $\mathrm{I}(\mathbf{r}, t, \Omega, \lambda)$, where $\mathbf{r}$ is the position vector indicating the location in space, t represents time, specifying when the measurement or observation is made, Ω is the solid angle vector, often expressed in spherical coordinates as (θ, φ). θ is the angle from the zenith (downward direction), and φ is the azimuthal angle around the zenith. λ stands for wavelength, indicating the specific wavelength of the light (or other electromagnetic radiation) being considered. The radiance is defined as the amount of radiant energy dW in the wavelength interval $[\lambda, \lambda + d\lambda]$ transported across an element of area dS into a solid angle $d\Omega$

Analytical Methods in Radiative Transfer, First Edition. Alexander Kokhanovsky, Vijay Natraj and Dmitry Efremenko.

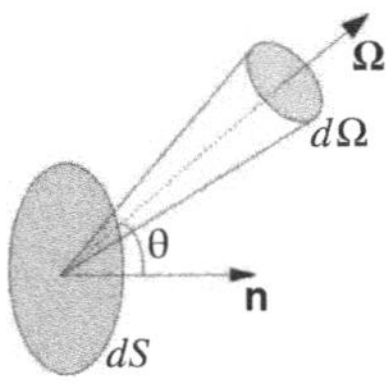

Figure 2.1 Schematic showing the definition of specific intensity.

oriented at an angle θ relative to the normal n of the area, within a time interval dt, as shown in Figure 2.1. It follows:

$$I = \frac{dW}{\cos\theta d\Omega d\lambda dS dt}. \tag{2.1}$$

Radiance is a function with six dimensions: three spatial dimensions, two directional dimensions, and one frequency (or wavelength) dimension. In the nonstationary case, adding a temporal dimension would extend it to seven dimensions, although this aspect will not be our focus. The inherent complexity of radiative transfer is hinted at by this multidimensionality, where even the mere storage of the intensity field in computer memory for an extended area presents a significant challenge due to its high dimensionality and mathematic complexity.

2.2 Formulation of the RTE

The change of light intensity during its propagation along the infinitesimal path dz in an absorbing medium in the absence of scattering effects is proportional to its value (at least as far as linear optics is concerned), as illustrated in Figure 2.2:

$$dI = -k_{\text{abs}} I d\ell, \tag{2.2}$$

where k_{abs} is the absorption coefficient at the wavelength λ. Integrating Eq. (2.2) under the assumption that k_{abs} does not change with distance ℓ, one can derive:

$$\mathrm{I} = \mathrm{I}_0 \exp(-k_{\text{abs}}\ell), \tag{2.3}$$

where $\mathrm{I}_0 = \mathrm{I}(z = 0)$.

Equation (2.3) is of fundamental importance for absorption spectroscopy. In particular, one can use Eq. (2.3) to derive the absorption coefficient k_{abs} at a given wavelength λ. Spectroscopy is a fundamental exploratory tool in the fields of chemistry, physics, and astronomy, allowing for the composition, physical, and electronic structure of matter to be investigated. Spectroscopy is used in physical and analytical chemistry because atoms and molecules have unique spectra. Consequently, these spectra can be used to detect, identify, and quantify information about the atoms and molecules. In particular, Kirchhoff and Bunsen (1861) have been able to establish the linkage between chemical elements and their unique spectral patterns.

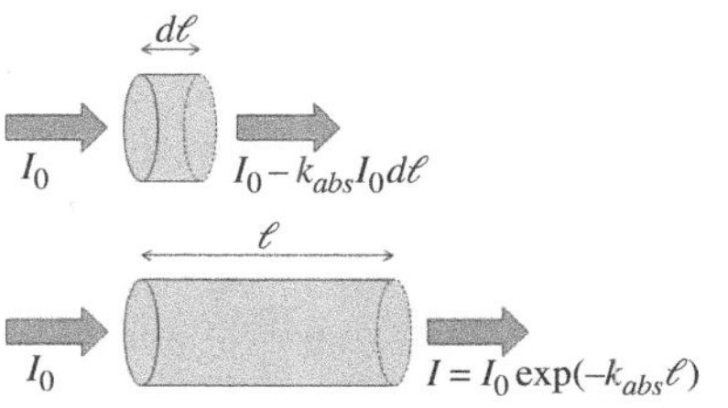

Figure 2.2 Absorption of light intensity while passing through a medium.

Let us imagine that light propagates in a turbid medium such as, for instance, a cloud in the sky. In this case, the change of light intensity along the infinitesimal path dz in the direction specified by the unit vector $\vec{n}$ can be represented as:

$$dI(\vec{n}) = -k_{ext}I(\vec{n})d\ell + \int_{4\pi} k_{sca}^{dif}(\vec{n}' \to \vec{n})I(\vec{n}')d\vec{n}'d\ell + j(\vec{n}), \tag{2.4}$$

where the second term accounts for light scattering from all other directions $\vec{n}'$ to a given direction of light propagation $\vec{n} = \sin\vartheta\cos\varphi\vec{x} + \sin\vartheta\sin\varphi\vec{y} + \cos\vartheta\vec{z}$, which is characterized by the zenith (ϑ) and azimuthal (φ) angles; j is the emission coefficient and accounts for the emission by the medium, while $\vec{x}$,$\vec{y}$,$\vec{z}$ are unit vectors along a 3D Cartesian coordinate system.

Let us introduce the cosine of the scattering angle: $\cos\Theta = \vec{n}\vec{n}'$. Then it follows using the definition of the scalar product for the cosine of the scattering angle from the direction specified by the angles (ϑ,φ) to the direction specified by the angles (ϑ',φ'):

$$\cos\Theta = \mu\mu' + \sqrt{(1-\mu^2)(1-\mu'^2)}\cos(\varphi - \varphi'), \tag{2.5}$$

where $\mu = \cos\vartheta$, $\mu' = \cos\vartheta'$.

Equation (2.4) follows from the energy conservation law as applied to a unit volume of a turbid medium. The parameter k_{ext} is called the extinction coefficient, and $k_{sca}^{dif}(\vec{n}' \to \vec{n})$ is called the differential scattering coefficient. It follows that

$$k_{ext} = k_{abs} + k_{sca}, \tag{2.6}$$

where

$$k_{sca} = 2\pi\int_0^{\pi} k_{sca}^{dif}(\Theta)\sin\Theta d\Theta \tag{2.7}$$

is the scattering coefficient, and we have assumed that the differential scattering coefficient depends just on the scattering angle Θ between the incident and scattered light directions. The differential scattering coefficient is related to the single-scattering phase function p as follows (see Section 2.5 for more details):

$$p(\Theta) = 4\pi\frac{k_{sca}^{dif}(\Theta)}{k_{sca}}. \tag{2.8}$$

Equation (2.4) is valid in the linear optics approximation under the assumption that there are no internal sources of radiation in the medium (e.g. thermal emission, laser-induced radiation sources, etc.). Also, possible changes in light frequency (e.g. fluorescence and luminescence) are ignored. Both radiation sources and media are assumed to be stationary. Therefore, changes in light intensity with time are not considered. Equation (2.4) can be considered as the generalization of Eq. (2.2) for the case of light scattering media. In particular, Eq. (2.2) follows from Eq. (2.4) under the assumption that there is no light scattering in the medium ($k_{\text{sca}}^{\text{dif}} = 0$).

Therefore, we conclude that the intensity of light propagating in a light scattering and absorbing medium is satisfied by the following integro-differential RTE:

$$\frac{d\mathrm{I}(\vec{n})}{d\ell} - \int_{4\pi} k_{\text{sca}}^{\text{dif}}(\vec{n}' \to \vec{n})\mathrm{I}(\vec{n}')d\vec{n}' - j(\vec{n}) + k_{\text{ext}}\mathrm{I}(\vec{n}) = 0, \tag{2.9}$$

where $\frac{d\mathrm{I}(\vec{n})}{d\ell}$ is the directional derivative. The direct problem of RTT is aimed at the determination of the spectral and angular distribution of light intensity I given by Eq. (2.9) inside turbid media and also at their boundaries for known (or assumed) local optical parameters ($k_{\text{sca}}^{\text{dif}}, k_{\text{ext}}$). This is the main subject of this book. The inverse RTT problem is related to the determination of the local optical properties of turbid media from angular and spectral distribution of light intensity I.

Let us rearrange the terms in Eq. (2.9) as follows:

$$-\frac{d\mathrm{I}(\vec{n})}{k_{\text{ext}}d\ell} = \mathrm{I}(\vec{n}) - S(\vec{n}), \tag{2.10}$$

where

$$S(\vec{n}) = \int_{4\pi} k_{\text{sca}}^{\text{dif}}(\vec{n}' \to \vec{n})\mathrm{I}(\vec{n}')d\vec{n}' + j(\vec{n}). \tag{2.11}$$

Equation (2.10) can be formally integrated along a chosen direction q, leading us to the following result:

$$\mathrm{I}(\vec{n}, q) = \mathrm{I}(\vec{n}, 0)\exp(-k_{\text{ext}}q) + \int_0^q S(\vec{n}, q')\exp[-k_{\text{ext}}(q' - q)]\,k_{\text{ext}}dq'. \tag{2.12}$$

This equation shows that the radiance at a given point q in the direction $\vec{n}$ consists of two parts. The first one corresponds to the Beer–Lambert law (see Eq. (2.3)), while the second one describes the influence of internal sources and its gradual attenuation as radiance propagates toward the specified coordinate. Equation (2.12) is referred to as the formal solution of the RTE ("formal" because S incorporates unknown I). This is illustrated in Figure 2.3. As q increases, the contribution of the first term incorporating the radiance at the origin exponentially decreases. At large distances, the impact of the second term may become dominant.

From Eq. (2.9), we may derive RTEs for plane-parallel, spherical, and cylindrical symmetries. Indeed, in the plane-parallel case, as shown in Figure 2.4, the coordinate z is related to the distance ℓ as

$$z = \mu\ell, \tag{2.13}$$

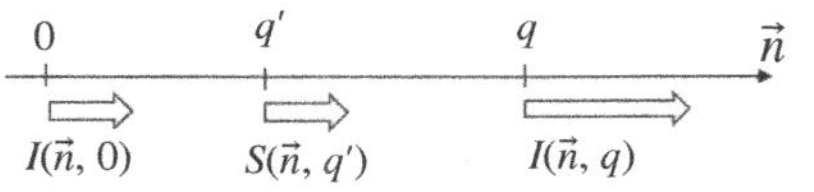

Figure 2.3 Illustration of the formal solution of the RTE corresponding to Eq. (2.12).

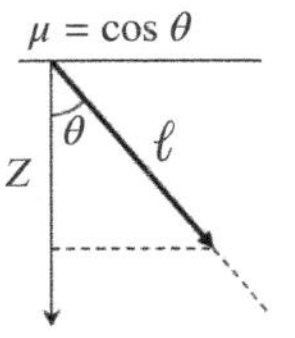

Figure 2.4 Illustration of one-dimensional geometry.

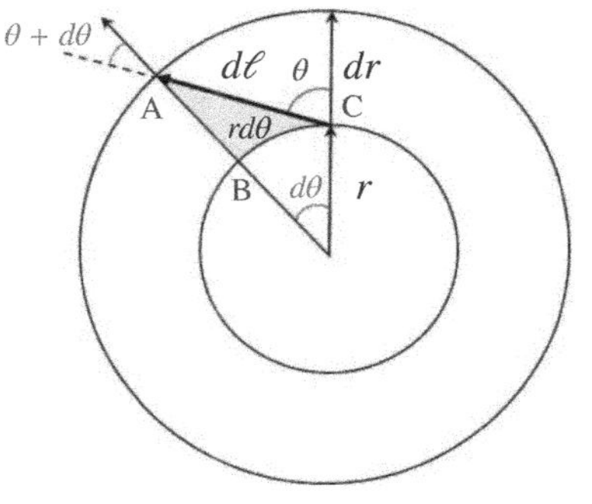

Figure 2.5 Illustration of spherical symmetry.

where $\mu = \cos\theta$, and θ is the angle between the vertical axis and the path. Substituting Eq. (2.13) into Eq. (2.9), we obtain

$$\mu\frac{d\mathrm{I}(\vec{n})}{dz} - \int_{4\pi} k_{\mathrm{sca}}^{\mathrm{dif}}(\vec{n}' \to \vec{n})\mathrm{I}(\vec{n}')d\vec{n}' - j(\vec{n}) + k_{\mathrm{ext}}\mathrm{I}(\vec{n}) = 0. \tag{2.14}$$

In the case of spherical symmetry, the radiance changes with the radius vector r and the angle θ between the radius vector and the light path. Consequently, the derivative $d\mathrm{I}/d\ell$ becomes

$$\frac{d\mathrm{I}}{d\ell} = \frac{d\mathrm{I}}{dr}\frac{dr}{d\ell} + \frac{d\mathrm{I}}{d\theta}\frac{d\theta}{d\ell}. \tag{2.15}$$

In Figure 2.5, we can see that in the triangle ABC:

$$\mathrm{AB} = \mathrm{AC}\cos(\theta + d\theta) \tag{2.16}$$

or

$$dr = d\ell\cos(\theta + d\theta) \approx d\ell\cos\theta. \tag{2.17}$$

Additionally, the arc of a circle BC is approximately equal to the side BC of triangle ABC, leading us to the following expression:

$$rd\theta = -\sin\theta d\ell. \tag{2.18}$$

Substituting Eqs. (2.15)–(2.18) into Eq. (2.9) we obtain:

$$\cos\theta\frac{\partial I}{\partial r}-\frac{\sin\theta}{r}\frac{\partial I}{\partial\theta}-\int_{4\pi}k_{\text{sca}}^{\text{dif}}(\vec{n}'\to\vec{n})\mathrm{I}(\vec{n}')d\vec{n}'+k_{\text{ext}}\mathrm{I}(\vec{n})-j(\vec{n})=0. \tag{2.19}$$

Taking into account that $\mu=\cos\theta$ and $d\mu=-\sin\theta d\theta$, this equation becomes

$$\mu\frac{\partial I}{\partial r}+\frac{1-\mu^2}{r}\frac{\partial I}{\partial\mu}-\int_{4\pi}k_{\text{sca}}^{\text{dif}}(\vec{n}'\to\vec{n})\mathrm{I}(\vec{n}')d\vec{n}'+k_{\text{ext}}\mathrm{I}(\vec{n})-j(\vec{n})=0. \tag{2.20}$$

In a three-dimensional Cartesian system, the RTE is given by

$$\xi\frac{d\mathrm{I}(\vec{n})}{dx}+\eta\frac{d\mathrm{I}(\vec{n})}{dy}+\mu\frac{d\mathrm{I}(\vec{n})}{dz}-\int_{4\pi}k_{\text{sca}}^{\text{dif}}(\vec{n}'\to\vec{n})\mathrm{I}(\vec{n}')d\vec{n}'-j(\vec{n})+k_{\text{ext}}\mathrm{I}(\vec{n})=0, \tag{2.21}$$

where ξ and η are the directional cosines of the direction $\vec{n}$ with respect to the x- and y-axes, respectively. The directional cosines are related to each other by the following relation:

$$\xi+\eta+\mu=1. \tag{2.22}$$

2.3 RTE with Thermal Source

Thermal emission refers to the phenomenon where all matter with a temperature greater than absolute zero emits electromagnetic waves. Previously, we have already considered Planck's function for describing the thermal emission. Now, we consider the RTE with the thermal source.

We start with recalling Kirchhoff's law of thermal radiation, often referred to as Kirchhoff's law of emission. It is a fundamental principle in the field of thermal radiation and thermodynamics. It states that for a body at thermal equilibrium, the emissivity (the efficiency of an object in emitting energy as thermal radiation) of a material at a given wavelength and temperature equals its absorptivity (the ability of the material to absorb radiation). In simpler terms, the ratio of the emissive power of a material to that of a blackbody (an idealized physical body that absorbs all incident electromagnetic radiation) at the same temperature is equal to the absorption coefficient of the material for radiation of the same wavelength.

Mathematically, Kirchhoff's law can be expressed as:

$$\epsilon(\lambda,T)=\alpha(\lambda,T), \tag{2.23}$$

where $\epsilon(\lambda,T)$ is the emissivity of the material at wavelength λ and temperature T, while $\alpha(\lambda,T)$ is the absorptivity of the material at the same wavelength and temperature.

Consequently, the thermal source can be described as follows:

$$j(\lambda,T)=\alpha(\lambda,T)B(\lambda,T). \tag{2.24}$$

Substituting Eq. (2.24) to Eq. (2.9), we get the RTE with the thermal source. It is important to note that this source term is isotropic, meaning that it emits radiation uniformly in all directions.

2.4 Optical Thickness and Single-scattering Albedo

Optical thickness and single-scattering albedo are two fundamental quantities in the study of radiative transfer, each playing a crucial role in understanding how light interacts with a medium.

The optical thickness τ measures the extent to which a medium absorbs and scatters light, rather than allowing it to pass through directly. It is a dimensionless quantity that quantifies the opacity of the medium. A high optical thickness means that the medium is highly absorptive or scattering, making it difficult for radiation to penetrate. Conversely, a low optical thickness indicates that the medium is relatively transparent. The optical thickness is determined by the properties of the medium (such as the concentration of absorbing or scattering particles) and the path length through the medium. It is integral to calculate the attenuation of light as it travels through a medium, following Beer–Lambert's law.

If z represents the geometrical thickness, then the corresponding optical thickness is given by:

$$\tau = \int_0^z k_{\text{ext}}(z')dz'. \tag{2.25}$$

If the extinction coefficient is constant throughout the layer, then this expression simplifies to:

$$\tau = k_{\text{ext}}z. \tag{2.26}$$

The optical thickness is a crucial parameter that significantly influences the behavior of the radiance field within a medium, effectively distinguishing between optically thin and optically thick environments. This distinction is important because the radiative properties and the character of light propagation vary markedly based on the optical thickness of the medium.

An optically thin medium does not significantly absorb or scatter light. As a result, radiation can travel through such a medium with minimal interaction, leading to a predominance of direct transmission over scattering and absorption. This characteristic allows for more straightforward radiative transfer calculations and simpler models of light propagation. Conversely, in an optically thick medium, there is substantial absorption and scattering of light. In these conditions, the radiation that passes through the medium undergoes multiple scattering events and absorption processes before it can emerge. This results in a more diffuse radiance field where the direct transmission is greatly diminished, and the light propagation is predominantly governed by scattering processes. Typical values of optical thicknesses of clouds depending on their types are listed in Table 2.1.

Single-scattering albedo, denoted by ω_0, is a measure of the scattering efficiency of a medium relative to its absorption and scattering combined. It is defined as the ratio of the scattering coefficient (k_{sca}) to the extinction coefficient (k_{ext}),

$$\omega_0 = \frac{k_{\text{sca}}}{k_{\text{ext}}}. \tag{2.27}$$

The single-scattering albedo is a dimensionless quantity that ranges from 0 to 1, with a value of 0 indicating that all interactions lead to absorption (no scattering), and a value of 1 indicating that all interactions lead to scattering (no absorption).

Table 2.1 Optical thickness ranges for different cloud types.

Cloud type	Typical optical thickness range
Cirrus (high)	0.1–3.0
Altostratus (mid)	0.3–4.0
Stratus (low)	5.0–20.0
Cumulonimbus (vertical)	15.0–100.0+

2.5 Single-scattering Phase Function

The single-scattering phase function describes how light or other electromagnetic radiation is scattered by a single particle or a single scattering event. It quantifies the angular distribution of scattered radiation relative to the direction of incident radiation. Essentially, the phase function indicates the probability of light being scattered into different directions after interacting with a particle.

It is commonly assumed that scattering is axially symmetric around the direction of incidence, which is true for spherical particles and also for randomly oriented nonspherical scatterers. As a result, the phase function p is a function of the scattering angle (or its cosine), denoted as $p(\cos\Theta)$. The normalization condition for p is formulated in such a way that the integral of the phase function across the whole sphere should give 4π:

$$\int_{\text{entire sphere}} p(\cos\Theta)d\Omega = 4\pi, \tag{2.28}$$

or in spherical coordinates ($\mu = \cos\vartheta, \varphi$),

$$\int_0^{2\pi}\int_{-1}^{1} p(\mu,\varphi)d\mu d\varphi = 4\pi. \tag{2.29}$$

For the azimuthally symmetric case, the normalization for $p(\mu)$ is

$$\int_{-1}^{1} p(\mu)d\mu = 2. \tag{2.30}$$

The phase function can be expanded into a series of Legendre polynomials P_n as:

$$p(\cos\Theta) = \sum_{n=0}^{\infty}(2n+1)\chi_n P_n(\cos\Theta), \tag{2.31}$$

where n denotes the order, and χ_n are the expansion coefficients, evaluated by considering a finite number of terms $N_{\max}$. The properties of Legendre polynomials are summarized in Appendix A. In particular, multiplying both sides of Eq. (2.31) by $P_{n'}(\cos\Theta)$, integrating from -1 to 1 over $\cos\Theta$, and using orthogonality property (A.4), we can express the expansion coefficients:

$$\chi_n = \frac{1}{2}\int_{-1}^{1} p(\cos\Theta)P_n(\cos\Theta)d\cos\Theta. \tag{2.32}$$

It is important to note that the first expansion coefficient (χ_1) in Eq. (2.31) is referred to as the asymmetry factor and denoted as g. Since $P_1(\cos\Theta) = \cos\Theta$ (see Table A.1), from Eq. (2.33) the asymmetry parameter can be expressed as

$$g = \frac{1}{2}\int_{-1}^{1} p(\cos\Theta)\cos\Theta d\cos\Theta. \tag{2.33}$$

and can be interpreted as the average cosine of the scattering angle, encapsulating the directionality of the phase function's scattering pattern. Values of g greater than zero suggest a bias toward forward scattering, indicating that the phase function is forward-peaked. Conversely, negative values of g imply a propensity for backscattering. Specifically, a value of $g = 1$ denotes exclusive forward scattering, whereas $g = -1$ signifies exclusive backscattering.

The Rayleigh phase function describes the angular distribution of light scattered by particles significantly smaller than the wavelength of the incident light, such as atmospheric molecules. This phenomenon is responsible for the sky's blue color due to the preferential scattering of shorter wavelengths of sunlight.

The mathematical expression for the Rayleigh phase function $P(\Theta)$ in terms of the scattering angle Θ for unpolarized light is given by:

$$P(\cos\Theta) = \frac{3}{4}(1 + \cos^2\Theta). \tag{2.34}$$

This formula demonstrates that the intensity of scattered light varies with the cosine squared of the scattering angle, indicating that light is most strongly scattered in the forward ($\Theta = 0$) and backward ($\Theta = \pi$) directions, and least strongly at angles perpendicular to the direction of incident light ($\Theta = \frac{\pi}{2}$).

The Mie phase function describes the angular distribution of light scattered by spherical particles of arbitrary sizes (Lorentz, 1890; Mie, 1908; Debye, 1909). Unlike Rayleigh scattering, which applies to much smaller particles, Mie scattering can be used to study light scattering by a wide range of particles, including those found in clouds, mists, and atmospheric aerosol. The Mie phase function is derived from Mie theory, a comprehensive solution to Maxwell's equations for the scattering of electromagnetic waves by spherical particles.

The resulting Mie phase function, which describes the angular distribution of scattered light, is significantly more complex than that for Rayleigh scattering and depends on both the size and the refractive index of the scattering particles relative to that of the surrounding medium.

The mathematical representation of the Mie phase function is not as straightforward as that for Rayleigh scattering, due to its dependence on a series of Bessel functions and Legendre polynomials to account for the various ways light can interact with particles of arbitrary sizes. Mie theory is summarized in Appendix B. For more insights into Mie theory, readers are encouraged to refer to review articles (Wriedt, 2012; Frisvad and Kragh, 2019). It is important to recognize that the Mie phase function for polydisperse particles can exhibit a complex shape, potentially featuring additional peaks in the backward hemisphere. This complexity arises from the diversity in particle sizes and compositions, which significantly influences the scattering pattern, including enhancements in certain directions that are not as pronounced in monodisperse systems.

Various analytical phase functions have been proposed to accurately reflect the scattering properties of particles, as outlined in Hapke (2009). Among these, the isotropic phase function stands as the simplest:

$$p(\cos\Theta) = 1. \tag{2.35}$$

For the isotropic phase function, the expansion coefficient χ_0 is equal to 1, with all subsequent coefficients equal to zero. Despite its simplicity and convenience for analytical derivations, this phase function lacks realism for atmospheric remote sensing applications. However, this function can be used in the framework of the so-called truncation method, when the phase function is represented as the combination of the highly extended in the forward direction Dirac delta function and the isotropic phase function (Dombrovsky and Baillis, 2010).

A linear approximation often employed in theoretical radiative transfer is given by:

$$p(\cos\Theta) = 1 + b\cos\Theta, \tag{2.36}$$

where b is a constant. This formulation simplifies radiative transfer calculations compared to anisotropic scattering phase functions and, in some instances, yields sufficiently accurate results for engineering purposes, as noted in Sobolev (1975).

In contrast, the needle phase function, which models exclusively forward scattering, is defined as:

$$p(\cos\Theta) = 2\delta(1 - \cos\Theta), \tag{2.37}$$

indicating scattering solely in the forward direction.

One widely used non-isotropic phase function is the Henyey–Greenstein function (Henyey and Greenstein, 1941):

$$p_{\mathrm{HG}}(g, \cos\Theta) = \frac{1 - g^2}{(1 + g^2 - 2g\cos\Theta)^{3/2}}. \tag{2.38}$$

This function, empirically derived to model light scattering by cosmic dust, serves to parameterize real scattering behaviors. Remarkably, the expansion coefficients of the Henyey–Greenstein phase function have the following simple form:

$$\chi_n = g^n. \tag{2.39}$$

The forward to backward scattering ratio is thus:

$$\frac{p_{\mathrm{HG}}(g, 1)}{p_{\mathrm{HG}}(g, -1)} = \left(\frac{1 + g}{1 - g}\right)^3. \tag{2.40}$$

For values of g approaching 1, the function is predominantly forward-peaked, whereas negative g values suggest dominant backscattering. The special case of $g = 0$ corresponds to isotropic scattering.

To enhance the description of backward scattering, a double Henyey–Greenstein function can be employed:

$$p(\alpha, g_1, g_2, \cos\Theta) = \alpha p_{\mathrm{HG}}(g_1, \cos\Theta) + (1 - \alpha)p_{\mathrm{HG}}(g_2, \cos\Theta), \tag{2.41}$$

where α acts as a weighting factor between two Henyey–Greenstein functions with asymmetry factors g_1 and g_2. An alternative modification, proposed by Cornette and Shanks (1992), is given by:

$$p(g, \cos\Theta) = \frac{3}{2}\frac{1-g^2}{2+g^2}\frac{1+g^2}{(1+g^2-2g\cos\Theta)^{3/2}}. \tag{2.42}$$

The expansion coefficients for this function are provided analytically, allowing for a more realistic representation of phase functions, especially for small particles.

Draine's phase function is a two-parameter distribution that generalizes the Cornette–Shanks phase function (Jendersie and d'Eon, 2023):

$$p_{CS}(g, \cos\Theta) = \frac{1-g^2}{(1+g^2-2g\cos\Theta)^{3/2}}\frac{1+\alpha\cos^2\Theta}{1+\alpha(1+2g^2)/3}. \tag{2.43}$$

Additionally, Kagiwada and Kalaba (1967) introduced the following phase function:

$$p(\cos\Theta) = \frac{2}{(b-\cos\Theta)\ln\frac{b+1}{b-1}}, \tag{2.44}$$

with the parameter b defined as:

$$b = \frac{r+1}{r-1}, \tag{2.45}$$

where r is the ratio of forward to backward scattering, typically chosen to model realistic cloud scattering with r around 2000 (Hunt, 1971). Note that Eqs. (2.43)–(2.44) may not satisfy normalization condition (Eq. (2.28)) and hence should be numerically normalized.

2.6 Orders of Scattering

If the original direction and the direction after scattering are described by two angles for each direction (i.e. μ, φ and μ', φ'), the phase function can be expressed in spherical coordinates. In this case, the normalization condition reads as follows:

$$\int_0^{2\pi}\int_{-1}^{1} p(\mu, \varphi, \mu', \varphi')d\mu' d\varphi = 4\pi. \tag{2.46}$$

In most cases, we may assume that the phase function does not depend on φ and φ', but rather $\varphi - \varphi'$, i.e.

$$p(\cos\Theta) = p(\mu, \varphi, \mu', \varphi') = p(\mu, \mu', \varphi - \varphi'). \tag{2.47}$$

The Legendre addition theorem states that for $\cos\Theta$ defined as in Eq. (2.5), the following representation for $P_n(\cos\Theta)$ is valid:

$$P_n(\cos\Theta) = \sum_{m=0}^{n}(2-\delta_{m0})\frac{(n-m)!}{(n+m)!}P_n^m(\mu)P_n^m(\mu')\cos[m(\varphi-\varphi')] \tag{2.48}$$

where P_n^m are the associated Legendre polynomials. It is convenient to expand $p(\mu, \mu', \varphi - \varphi')$ using the addition theorem of Legendre polynomials in Eq. (2.31):

$$p(\mu, \mu', \varphi - \varphi') = \sum_{m=0}^{n} (2 - \delta_{m0}) \sum_{n=m}^{\infty} (2n+1) \frac{(n-m)!}{(n+m)!} \chi_n P_n^m(\mu) P_n^m(\mu') \cos[m(\varphi - \varphi')]. \tag{2.49}$$

Introducing the normalized associated Legendre polynomials as

$$\bar{P}_n^m(\mu) = \sqrt{(2n+1)\frac{(n-m)!}{(n+m)!}} P_n^m(\mu), \tag{2.50}$$

Eq. (2.49) can be written in a more compact form:

$$p(\mu, \mu', \varphi - \varphi') = \sum_{m=0}^{n} (2 - \delta_{m0}) \sum_{n=m}^{\infty} \chi_n \bar{P}_n^m(\mu) \bar{P}_n^m(\mu') \cos[m(\varphi - \varphi')]. \tag{2.51}$$

In practice, it is preferable to use the normalized polynomials since their computations are computationally more stable (Dave and Armstrong, 1970).

Let us consider two consecutive scattering events. The angular distribution of double scattering radiation is given by

$$\bar{p}^2(\mu_0, \mu, \varphi) = \int_0^{2\pi} \int_{-1}^{1} p(\mu_0, \mu', \varphi') p(\mu', \mu, \varphi - \varphi') \, d\mu' d\varphi'. \tag{2.52}$$

Substituting Eq. (2.51) into Eq. (2.52) and applying the orthogonality property (A.6), we obtain:

$$\bar{p}^2(\mu_0, \mu, \varphi) = \sum_{m=0}^{M_{\max}} (2 - \delta_{m0}) \sum_{n=m}^{N_{\max}} \chi_n^2 \bar{P}_n^m(\mu_0) \bar{P}_n^m(\mu) \cos(m\varphi). \tag{2.53}$$

Repeating this procedure for $\bar{p}^s$, we get an expression for the angular distribution of the radiation after s scattering events:

$$\bar{p}^s(\mu_0, \mu, \varphi) = \sum_{m=0}^{M_{\max}} (2 - \delta_{m0}) \sum_{n=m}^{N_{\max}} \chi_n^s \bar{P}_n^m(\mu_0) \bar{P}_n^m(\mu) \cos(m\varphi). \tag{2.54}$$

Note that $\{\chi_n^s\}_{n=0,1,2...}$ becomes smaller as the order of scattering s increases. In the limit $s \to \infty$, only $\chi_0^s = 1$, while the remaining expansion coefficients vanish. Hence p^s becomes isotropic.

The angular characteristics of the scattered radiance distribution are primarily the result of low-order scattering events, whereas the "background" radiation originates from higher-order scattering. It is important to note that after several scattering events, photons tend to lose memory of their original source, causing the radiation to reach a state of diffusion equilibrium. This phenomenon was utilized by Sobolev (1943) in his development of an approximate method for solving the problem of anisotropic scattering of monochromatic radiation (see the comprehensive review by Nagirner (2016) for more details).

2.7 One-dimensional RTE

Let us assume that the properties of a turbid medium vary only along the vertical coordinate z directed toward the interior of the medium. We shall assume that the origin of the coordinate system is at the top of a plane-parallel layer illuminated by a wide unidirectional light beam with flux πF_0. In this case, the intensity I depends only on the coordinate z and the observation direction specified by the zenith ϑ and the azimuthal angle φ. For now, we neglect the emission of the medium. Then Eq. (2.7) can be transformed into the following form for plane-parallel turbid layers:

$$\mu \frac{d\mathrm{I}(\vartheta,\varphi)}{d\tau} - \frac{\omega_0}{4\pi}\int_0^{2\pi} d\varphi \int_0^{\pi} p(\Theta)\mathrm{I}(\vartheta',\varphi')\sin\Theta d\Theta \\ + \mathrm{I}(\vartheta,\varphi) - (1-\omega_0)B(\lambda,T) = 0, \tag{2.55}$$

where $\mu = \cos\vartheta$, $\mu' = \cos\vartheta'$, τ is the optical depth, ω_0 is the single-scattering albedo, Θ is the scattering angle defined as in Eq. (2.5), and

$$p(\Theta) = \frac{4\pi k_{\mathrm{sca}}^{\mathrm{dif}}(\Theta)}{k_{\mathrm{sca}}} \tag{2.56}$$

is the phase function normalized as follows (see Eq. (2.7)):

$$\frac{1}{2}\int_0^{\pi} p(\Theta)\sin\Theta d\Theta = 1. \tag{2.57}$$

Note that we are usually interested in finding the upwelling intensity at the top of the medium (i.e. reflected radiation) and the downwelling intensity at the bottom of the medium (i.e. transmitted radiation), while the upwelling intensity at the bottom and the downwelling intensity at the top are known and taken into account through the boundary conditions. In this regard, the function I in Eq. (2.55) explicitly depends on the spatial coordinate τ.

The intensity $I(\vartheta,\varphi)$ can be presented in the following form:

$$\mathrm{I}(\vartheta,\varphi) = I(\vartheta,\varphi) + I_{dir}\delta(\mu-\mu_0)\,\delta(\varphi-\varphi_0). \tag{2.58}$$

Here, δ is the Dirac delta function, $\mu_0 = \cos\vartheta_0$ is the cosine of the incidence zenith angle ϑ_0, and φ_0 is the azimuthal angle of the incident light. The first term represents the diffuse radiation, and the second term represents the direct radiation. One can derive, after the substitution of Eq. (2.58) in Eq. (2.55):

$$\cos\vartheta \frac{dI(\vartheta,\varphi)}{d\tau} - \frac{\omega_0}{4\pi}\int_0^{2\pi} d\varphi' \int_0^{\pi} d\vartheta' p(\Theta) I(\vartheta',\varphi') + I(\vartheta,\varphi) \\ = \frac{\omega_0}{4\pi} p(\Theta_s) F_0 \exp\left(-\frac{\tau}{\cos\vartheta_0}\right) + (1-\omega_0)B(\lambda,T), \tag{2.59}$$

$$I_{dir} = F_0 \exp\left(-\frac{\tau}{\cos\vartheta_0}\right), \tag{2.60}$$

where, as follows from spherical trigonometry:

$$\cos\Theta_s = \mu\mu_0 + \sqrt{(1-\mu_0^2)(1-\mu^2)}\cos(\varphi-\varphi_0). \tag{2.61}$$

Equation (2.60) gives the analytical result for the direct light intensity I_{dir}. The diffuse light intensity is governed by Eq. (2.59). On first glance, one may conclude that Eq. (2.59) is more complex as compared to Eq. (2.55) because it contains an additional term. However, as a matter of fact, the integro-differential Eq. (2.59) is easier to solve using numerical techniques because it does not contain the singularities present in Eq. (2.58). There are several numerical codes for the solution of the RTE.[1] As a matter of fact, there are several numerical methods to solve Eq. (2.59) to required accuracy, which can be found in various textbooks (see Stamnes et al. (2017), Peraiah (2001), Efremenko and Kokhanovsky (2021), and references therein). The focus of this book is on analytical methods in radiative transfer, with the aim of deriving approximate solutions of the integro-differential RTE representing various commonly occurring situations in remote sensing. Usually, analytical methods are faster than numerical methods while being accurate enough, which makes it possible to apply them for solving several engineering problems, including spectroscopy of turbid media. In addition, analytical methods are based on interesting physical ideas and assumptions, which allow one to look at RTT from a physical, rather than mathematical, point of view.

2.8 Formal Solution of the RTE

Let us apply Eq. (2.12) to a slab in order to derive a formal solution of the one-dimensional RTE. We consider a homogeneous layer of the optical thickness $\tau_{\rm L}$. As boundary conditions, we have the downwelling radiation on top $\mathrm{I}(0,\mu>0,\varphi)$ and the upwelling radiation at the bottom $\mathrm{I}(\tau_{\rm L},\mu<0,\varphi)$. The result reads as follows (Chandrasekhar, 1950):

$$\mathrm{I}(\tau,\mu>0,\varphi) = \mathrm{I}(0,\mu>0,\varphi)\,e^{-\frac{\tau}{\mu}} + \int_0^{\tau} S(\tau',\mu,\varphi)\,e^{-\frac{\tau-\tau'}{\mu}}\frac{d\tau'}{\mu} \tag{2.62}$$

for the downwelling radiation, and

$$\mathrm{I}(\tau,\mu<0,\varphi) = \mathrm{I}(\tau_{\rm L},\mu<0,\varphi)\,e^{-\frac{\tau_L-\tau}{|\mu|}} + \int_{\tau}^{\tau_{\rm L}} S(\tau',\mu,\varphi)\,e^{-\frac{\tau'-\tau}{|\mu|}}\frac{d\tau'}{|\mu|} \tag{2.63}$$

for the upwelling radiation. These solutions align with a clear physical concept, as depicted in Figure 2.6. The radiance reaching the slab's boundaries diminishes exponentially as it travels from the top or bottom of the layer to a specific coordinate τ. For downward radiation, the path is defined as τ/μ, whereas for upward radiation, the path is expressed as $(\tau_{\rm L}-\tau)/|\mu|$, utilizing the absolute value of μ because μ is negative for upward radiation. The radiation emitted by the source S at coordinate τ' experiences exponential attenuation as it moves through the medium from τ' to τ. It is crucial to consider the radiation sources at all interior points, adjusted by the corresponding exponential decay factors.

---

1 https://en.wikipedia.org/wiki/Atmospheric_radiative_transfer_codes

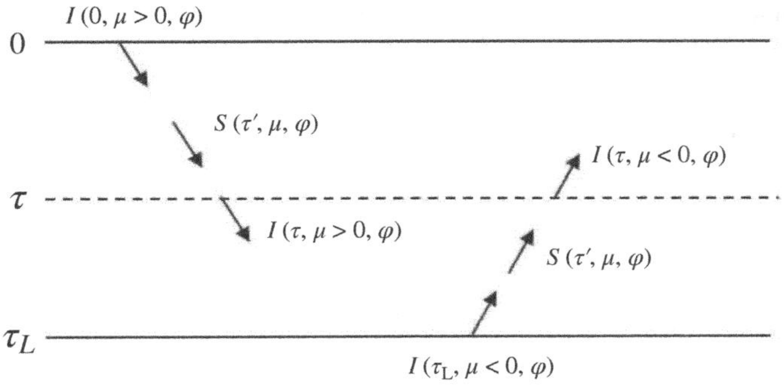

Figure 2.6 Attenuation of the radiation and internal sources. The downwelling radiation is formed by the external downwelling radiation, which originates from incident radiation at the top, and internal sources situated above a specified level. Conversely, the upwelling radiation is due to external upwelling radiation at the layer bottom and internal sources located beneath the chosen level τ.

For the radiation transmitted through the layer, we have

$$\mathrm{I}(\tau_{\mathrm{L}},\mu>0,\varphi)=\mathrm{I}(0,\mu>0,\varphi)e^{-\frac{\tau_{\mathrm{L}}}{\mu}}+\int_{0}^{\tau_{\mathrm{L}}}S(\tau',\mu,\varphi)e^{-\frac{\tau_{\mathrm{L}}-\tau'}{\mu}}\frac{d\tau'}{\mu},\tag{2.64}$$

while the reflected light intensity is given by

$$\mathrm{I}(0,\mu<0,\varphi)=\mathrm{I}(\tau_{\mathrm{L}},\mu<0,\varphi)e^{-\frac{\tau_{\mathrm{L}}}{|\mu|}}+\int_{0}^{\tau_{\mathrm{L}}}S(\tau',\mu,\varphi)e^{-\frac{\tau'}{|\mu|}}\frac{d\tau'}{|\mu|}.\tag{2.65}$$

We note again that the source S may be unknown *a priori* and be dependent on the radiance I.

2.9 RTE for Azimuthal Harmonics of the Intensity

As we can see in Eq. (2.55), the unknown function I depends on two angular variables. It is a common procedure to factor out the φ-dependency by considering the Fourier expansion of the total intensity and the single-scattering phase function. As these functions are even with respect to φ, the Fourier expansion turns into a cosine expansion. Thus, we have

$$\mathrm{I}(\vartheta,\varphi)=\sum_{m=0}^{\infty}\mathrm{I}_m(\vartheta)\cos m\varphi,\tag{2.66}$$

$$p(\Theta)=\sum_{m=0}^{\infty}(2-\delta_{m0})p_m(\mu,\mu')\cos[m(\varphi-\varphi')],\tag{2.67}$$

where δ is the Kronecker delta, i.e.

$$\delta_{ij}=\begin{cases}0 & \text{if } i\neq j,\\ 1 & \text{if } i=j,\end{cases}\tag{2.68}$$

while θ, μ, and μ' correspond to Eq. (2.5). The term $(2-\delta_{m0})$ will make the final expressions a bit more compact.

Let us consider the integral term of Eq. (2.55):

$$J(\vartheta,\varphi) = \frac{\omega_0}{4\pi}\int_0^{2\pi} d\varphi \int_0^{\pi} p(\Theta)\mathrm{I}(\vartheta,\varphi)\sin\Theta d\Theta. \tag{2.69}$$

Substituting Eqs. (2.66) and (2.67) into Eq. (2.69), we obtain

$$\begin{aligned} J(\vartheta,\varphi) = \frac{\omega_0}{4\pi}\int_0^{2\pi}\int_{-1}^{1}\Bigg\{\left[\sum_{m'=0}^{n}(2-\delta_{m'0})p_{m'}(\mu,\mu')\cos[m'(\varphi-\varphi')]\right] \\ \times\left[\sum_{m=0}^{\infty}\mathrm{I}_m(\mu')\cos m\varphi'\right]\Bigg\}d\mu' d\varphi'. \end{aligned} \tag{2.70}$$

Note that, to avoid confusion, the summation in the second sum is performed over m' rather than m. Interchanging integration and summation, we have

$$\begin{aligned} J(\vartheta,\varphi) = \frac{\omega_0}{4\pi}\sum_{m=0}^{\infty}\sum_{m'=0}^{\infty}\Bigg\{(2-\delta_{m'0})\int_0^{2\pi}\cos[m'(\varphi-\varphi')]\cos m\varphi' d\varphi' \\ \times\int_{-1}^{1}\mathrm{I}_m(\mu')p_{m'}(\mu,\mu')d\mu'\Bigg\}. \end{aligned} \tag{2.71}$$

By using trigonometry and the orthogonality property of cosine functions over the range $[0, 2\pi]$, one can prove that

$$\int_0^{2\pi}\cos m\varphi'\cos m'(\varphi-\varphi')d\varphi' = (1+\delta_{m0})\pi\cos m\varphi. \tag{2.72}$$

Substituting this result in Eq. (2.71) and noting that

$$(2-\delta_{m0})(1+\delta_{m0}) = 2, \tag{2.73}$$

we obtain:

$$J(\vartheta,\varphi) = \frac{\omega_0}{2}\sum_{m=0}^{\infty}\cos m\varphi\int_{-1}^{1}\mathrm{I}_m(\mu')p_m(\mu,\mu')d\mu'. \tag{2.74}$$

The azimuthal expansion of the thermal source is straightforward due to its isotropic nature, meaning it emits radiation equally in all directions and thus lacks dependence on the azimuth angle. Consequently, it contributes solely to the zeroth harmonic in the expansion. Keeping that in mind, substituting Eqs. (2.66) and (2.74) into Eq. (2.55) and applying the orthogonality property of cosine functions, we obtain:

$$\begin{aligned} \mu\frac{d\mathrm{I}_m(\vartheta)}{d\tau} - \frac{\omega_0}{2}\int_{-1}^{1}p_m(\mu,\mu')\mathrm{I}_m(\vartheta')d\mu' + \mathrm{I}_m(\vartheta) \\ -\delta_{m0}(1-\omega_0)B(\lambda,T) = 0. \end{aligned} \tag{2.75}$$

Note that in the case of the incidence of radiation along the normal to a scattering layer, and similarly, when observations are made normal to this layer, there is no azimuthal dependence for the horizontally homogeneous media and

$$\mathrm{I}(\vartheta,\varphi) = \mathrm{I}_0(\vartheta), \tag{2.76}$$

while $I_m(\vartheta)$ vanish for $m \geq 1$. As a matter of fact, $I_0(\vartheta)$ is the azimuthally averaged radiance. Also note that the irradiance F (the amount of radiant energy transferred across a unit area in unit time in unit wavelength interval) can be computed as follows:

$$F = \int_0^{2\pi} \int_0^{\pi/2} I(\vartheta, \varphi) \cos\vartheta \sin\vartheta d\vartheta d\varphi. \tag{2.77}$$

Substituting Eq. (2.66) into Eq. (2.77), we derive

$$F = 2\pi \int_0^{\pi/2} I_0(\vartheta) \cos\vartheta \sin\vartheta d\vartheta. \tag{2.78}$$

Thus, if we are interested only in irradiances (radiant fluxes), then we have to compute only $I_0(\vartheta)$ instead of $I(\vartheta, \varphi)$.

2.10 Radiance Moments

Moments of the radiance field in the context of radiative transfer refer to integral quantities that characterize the distribution of radiance (or radiant intensity) across a medium. These moments are derived from the RTE and provide a simplified description of the radiation field, capturing essential features such as the average intensity, the flux of radiation, and the anisotropy of the radiation field. The moments are particularly useful in applications where detailed spatial or angular distribution of radiance is less critical, or where such detailed information is not readily available.

Eq. (2.78) can be generalized by considering radiance moments:

$$M_n = 2\pi \int_{-\pi/2}^{\pi/2} I_0(\vartheta) \cos^n\vartheta \sin\vartheta d\vartheta. \tag{2.79}$$

In particular, for $n = 0$, we get the mean intensity:

$$P = 2\pi \int_{-\pi/2}^{\pi/2} I_0(\vartheta) \sin\vartheta d\vartheta. \tag{2.80}$$

For $n = 1$, we get the radiant flux (Eddington flux):

$$F = 2\pi \int_{-\pi/2}^{\pi/2} I_0(\vartheta) \cos\vartheta \sin\vartheta d\vartheta. \tag{2.81}$$

Finally, if $n = 2$, we obtain the so-called K-integral:

$$K = 2\pi \int_{-\pi/2}^{\pi/2} I_0(\vartheta) \cos^2\vartheta \sin\vartheta d\vartheta. \tag{2.82}$$

Let us consider Eq. (2.75) in a plane-parallel medium without absorption:

$$\mu \frac{dI_0(\vartheta)}{d\tau} - \frac{1}{2} \int_{-1}^{1} p_0(\mu, \mu') I_0(\vartheta') d\mu' + I_0(\vartheta) = 0. \tag{2.83}$$

Integrating all terms over $d\mu = \sin\vartheta d\vartheta$ and taking into account Eq. (2.57), we may see that the last two terms in Eq. (2.83) cancel out. The remaining first term gives

$$\frac{dF}{d\tau} = 0. \tag{2.84}$$

Thus, in a plane-parallel medium without absorption, the flux remains constant with the layer optical depth as one may expect on physical grounds. Similarly, multiplying Eq. (2.75) by μ and integrating over $d\mu = \sin\vartheta d\vartheta$, we obtain:

$$\frac{dK}{d\tau} = \frac{1}{4}(1-g)F, \tag{2.85}$$

where g is the asymmetry parameter. Here we take into account that μ is the first Legendre polynomial, as according to the addition theorem of Legendre polynomials and orthogonality property (see Appendix A), only p_1 related to g will survive during integration. If F is constant, this equation can be easily integrated:

$$K = \frac{1}{4}F[(1-g)\tau + C], \tag{2.86}$$

where C is a constant. Equation (2.86) is referred to as the K-integral.

Eqs. (2.84) and (2.86) are straightforward yet pivotal for validating more sophisticated solutions and assessing certain properties of the radiance field. These equations serve as benchmarks, enabling comparison with complex models to ensure accuracy and consistency in the analysis of radiative transfer phenomena.

2.11 Light Reflection from Layered Media

Consider the reflection and transmission of radiation from a two-layer system. Each layer is characterized by reflection and transmission coefficients, denoted as r and t, respectively. The reflection coefficient for the combined two-layer system can be derived by analyzing a sequence of reflections and transmissions within each layer.

The reflection from this two-layer system is illustrated in Figure 2.7. The total reflection coefficient r_{1+2} for the system can be expressed as the following series:

$$r_{1+2} = r_1 + t_1 r_2 t_1 + t_1 r_2 r_1 r_2 t_1 + t_1 r_2 r_1 r_2 r_1 r_2 t_1 + \ldots \tag{2.87}$$

By applying the formula for a decaying geometric progression, this expression simplifies to:

$$r_{1+2} = r_1 + \frac{r_2 t_1^2}{1 - r_1 r_2} \tag{2.88}$$

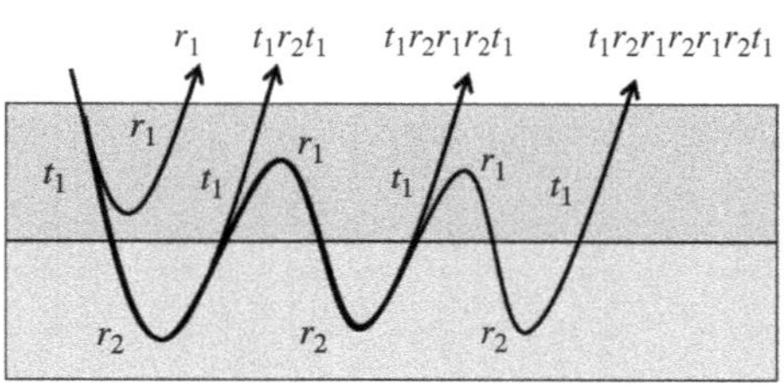

Figure 2.7 Schematic diagram of reflection from two slabs.

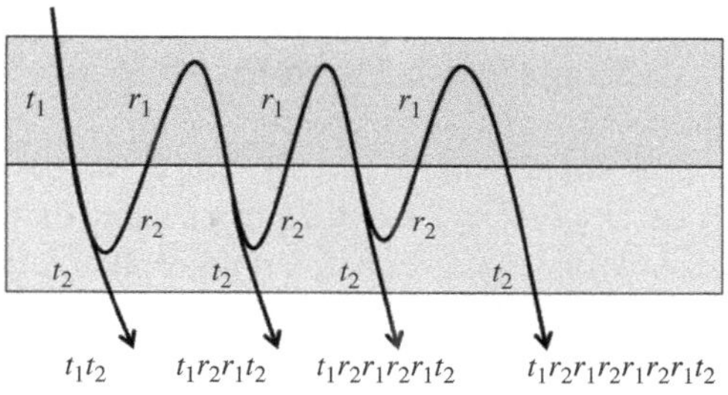

Figure 2.8 Schematic diagram of transmission from two slabs.

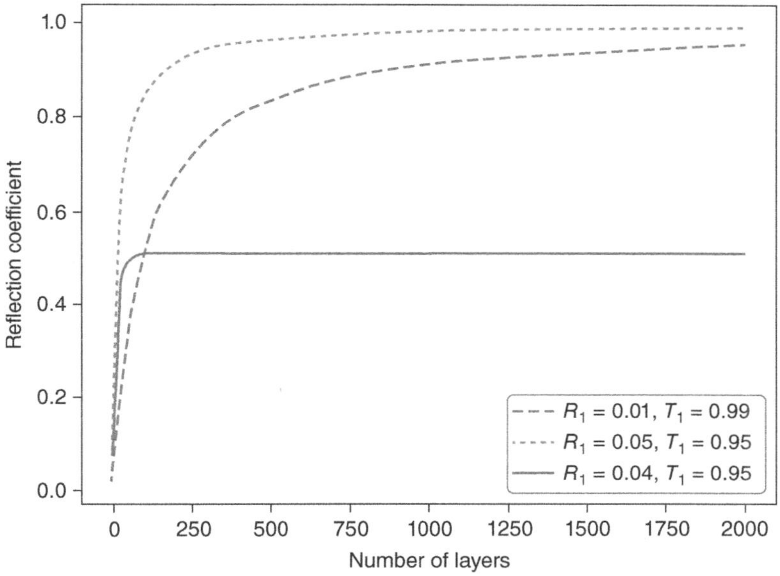

Figure 2.9 Reflection coefficient as a function of the number of layers.

Similarly, the transmission coefficient t_{1+2} involves a sequence of processes as illustrated in Figure 2.8. The corresponding expression is given by:

$$t_{1+2} = t_1t_2 + t_1r_2r_1t_2 + t_1r_2r_1r_2r_1t_2 + t_1r_2r_1r_2r_1r_2r_1t_2 + \dots \tag{2.89}$$

After applying the formula for geometric progression, this can be reduced to:

$$t_{1+2} = \frac{t_1t_2}{1 - r_1r_2}. \tag{2.90}$$

These relations were obtained by Stokes (1862).

Using obtained formulas, we can straightforwardly demonstrate the so-called invariance principle. To do this, we incrementally add layers, applying Eqs. (2.88) and (2.90) each time to update the reflection coefficient for the growing stack of layers. Each layer is characterized by a reflection coefficient r_1 and a transmission coefficient t_1. The overall reflection and transmission coefficients of the current stacked system before adding a layer are then r_2 and t_2, respectively, while r_{1+2} and t_{1+2} (see Eqs. (2.88) and (2.90)) are corresponding coefficients after adding a layer. Figure 2.9 illustrates how the reflection coefficient evolves as the number of layers in the stack increases. Notably, the curves reach plateaus, indicating that adding additional layers to a sufficiently thick stack does not further alter the reflection properties.

One curve (where $t_1 = 0.95$ and $r_1 = 0.04$) corresponds to cases with absorption in the layer, as $t_1 + r_1 < 1$, resulting in a plateau that is lower than those observed in cases without absorption. For cases without absorption, the curves asymptotically approach a reflection coefficient of 1 as the number of layers tends toward infinity.

3

Optically Thin Media and Media with Strongly Anisotropic Scattering

3.1 Single-scattering Approximation

The simplest approximation that can be used to solve the radiative transfer equation (RTE) is the single-scattering approximation. It can be used for media with small optical thickness $\tau_0 = k_{\text{ext}}h$, where h is the geometrical thickness of the medium. The trajectories of photons reflected by a medium that has experienced single- and multiple-scattering processes are shown in Figure 3.1. One can see that in the single-scattering approximation, the photon can be scattered only once. The concept of the single-scattering approximation in the field of wave propagation in random media was studied by Ishimaru (1978).

In this case, multiple scattering of photons represented by the integral term in Eq. (2.59) can be neglected, and the RTE is transformed to the following ordinary differential equation:

$$\cos\vartheta \frac{dI(\tau,\vartheta,\varphi)}{d\tau} + I(\tau,\vartheta,\varphi) = \frac{\omega_0}{4\pi} p(\Theta_s) F_0 \exp\left(-\frac{\tau}{\cos\vartheta_0}\right). \tag{3.1}$$

Let us consider the radiance as a function of optical depth, denoted by τ. Specifically, we are interested in the dependence of radiance on τ, expressed as $I(\tau)$. For clarity, we omit the explicit dependence of radiance I on angular variables in this discussion. Let us represent Eq. (3.1) in the following form:

$$\frac{dI(x)}{dx} + I(x) = \frac{1}{4\pi}\omega_0(x) p(x) F_0 \exp(-sx), \tag{3.2}$$

where $x = \tau / \cos\vartheta$, $s = \cos\vartheta / \cos\vartheta_0$ and we have omitted other arguments for simplicity. It is assumed that both single-scattering albedo and phase function depend on the optical depth. This equation can be solved analytically. For this, we multiply both sides of Eq. (3.2) by e^x. Then it follows:

$$\exp(x)\frac{dI(x)}{dx} + \exp(x)I(x) = \frac{\omega_0(x)}{4\pi} p(x) F_0 \exp(-(s-1)x) \tag{3.3}$$

or

$$\frac{d[I_d(x)e^x]}{dx} = \frac{\omega_0(x)}{4\pi} p(x) F_0 \exp(-(s-1)x). \tag{3.4}$$

Analytical Methods in Radiative Transfer, First Edition. Alexander Kokhanovsky, Vijay Natraj and Dmitry Efremenko.

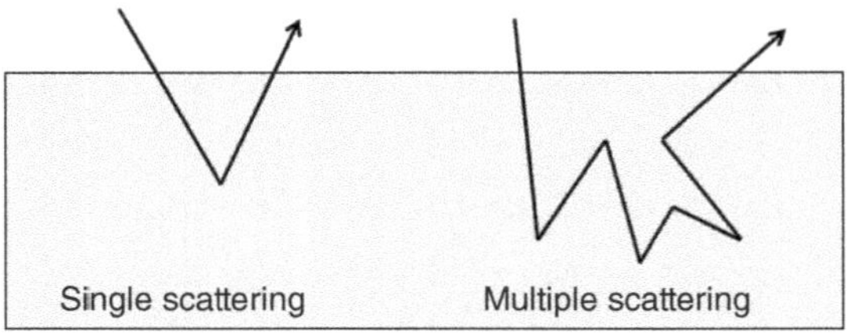

Figure 3.1 Trajectories of photons experiencing single- and multiple-scattering processes.

It means that

$$I_{\mathrm{d}}(x) = \frac{F_0}{4\pi} e^{-x} \int_a^x \omega_0(x') p(x') \exp(-(s-1)x') dx'. \tag{3.5}$$

Let us apply boundary conditions assuming that there is no diffuse light coming from the top and bottom (black underlying surface) of the light scattering layer

$$\vec{I}_{\mathrm{d}\uparrow}(x=0) = 0, \ \cos\vartheta > 0, \tag{3.6}$$

$$\vec{I}_{\mathrm{d}\downarrow}(x=x_0) = 0, \ \cos\vartheta < 0. \tag{3.7}$$

It follows for the downwelling $\vec{I}_{\mathrm{d}\downarrow}(x)$ and upwelling $\vec{I}_{\mathrm{d}\uparrow}(x)$ intensities:

$$I_{\mathrm{d}\downarrow}(x) = \frac{F_0}{4\pi} e^{-x} \int_0^x \omega_0(x') p(x') \exp(-(s-1)x') dx', \tag{3.8}$$

$$I_{\mathrm{d}\uparrow}(x) = \frac{F_0}{4\pi} e^{-x} \int_{x_0}^x \omega_0(x') p(x') \exp(-(s-1)x') dx' \tag{3.9}$$

and the boundary conditions are satisfied automatically. A simple integration under the assumption of a vertically homogeneous turbid layer gives

$$I_{\mathrm{d}\downarrow}(x) = \frac{\omega_0 p(\Theta) F_0}{4\pi (s-1)} \{e^{-x} - e^{-sx}\}, \tag{3.10}$$

$$I_{\mathrm{d}\uparrow}(x) = \frac{\omega_0 p(\Theta) F_0}{4\pi (s-1)} \{e^{-x-(s-1)x_0} - e^{-sx}\} \tag{3.11}$$

for arbitrary x. It follows at the boundaries of a layer:

$$I_{\mathrm{t}} \equiv I_{\mathrm{d}\downarrow}(x=x_0) = \frac{\omega_0 p(\Theta) F_0}{4\pi (s-1)} \{e^{-x_0} - e^{-sx_0}\}, \tag{3.12}$$

$$I_{\mathrm{r}} = I_{\mathrm{d}\uparrow}(x=0) = \frac{\omega_0 p(\Theta) F_0}{4\pi (s-1)} \{e^{-(s-1)x_0} - 1\}, \tag{3.13}$$

or

$$I_{\mathrm{t}} = \frac{\xi \omega_0 p(\Theta) F_0}{4\pi (\xi - \eta)} \left\{\exp\left[-\frac{\tau_0}{\xi}\right] - \exp\left[-\frac{\tau_0}{\eta}\right]\right\}, \tag{3.14}$$

$$I_r = \frac{\xi\omega_0 p(\Theta) F_0}{4\pi(\xi+\eta)}\left\{1 - \exp\left[\left(\frac{1}{\xi}+\frac{1}{\eta}\right)\tau_0\right]\right\}, \tag{3.15}$$

where $\eta = |\cos\vartheta|$, $\xi = \cos\vartheta_0$, τ_0 is the optical thickness of the scattering layer. Let us introduce the transmission (T) and reflection (R) functions for a turbid layer:

$$T = \frac{\pi I_t}{F_0\mu_0}, \tag{3.16}$$

and

$$R = \frac{\pi I_r}{F_0\mu_0}. \tag{3.17}$$

Obviously, it follows

$$T = \frac{\omega_0 p(\Theta_s)}{4(\xi-\eta)}\left\{\exp\left[-\frac{\tau_0}{\xi}\right] - \exp\left[-\frac{\tau_0}{\eta}\right]\right\}, \tag{3.18}$$

$$R = \frac{\omega_0 p(\Theta_s)}{4(\xi+\eta)}\left\{1 - \exp\left[\left(\frac{1}{\xi}+\frac{1}{\eta}\right)\tau_0\right]\right\}, \tag{3.19}$$

For the semi-infinite medium, Eq. (3.19) becomes

$$R = \frac{\omega_0 p(\Theta_s)}{4(\xi+\eta)}. \tag{3.20}$$

Let us note that this result can be directly derived from Eq. (4.58) without solving Eq. (3.1). One derives for $\xi = \eta$:

$$T = \frac{\omega_0\tau_0 p(\Theta_s)}{4\xi^2}\exp\left[-\frac{\tau_0}{\xi}\right], \tag{3.21}$$

$$R = \frac{\omega_0 p(\Theta_s)}{8\xi}\left(1 - \exp(-2\tau_0/\xi)\right). \tag{3.22}$$

The cosine of the scattering angle is

$$\cos\Theta_s = -\eta\mu_0 + \sqrt{(1-\mu_0^2)(1-\eta^2)}\cos(\varphi-\varphi_0) \tag{3.23}$$

for reflected radiation and

$$\cos\Theta_s = \eta\mu_0 + \sqrt{(1-\mu_0^2)(1-\eta^2)}\cos(\varphi-\varphi_0) \tag{3.24}$$

for transmitted radiation.

The main benefit of the single-scattering approximation is its simplicity. Despite a quite strong assumption that the photon experiences only one scattering event, this approximation can be accurate in some cases. This is in particular true for optically thin scattering layers or when the single-scattering albedo is very small. Single scattering is also important when polarization is considered, as the polarization of multiple-scattered light is typically much lower than that of light scattered only once. A sensitivity study based on the successive orders of scattering method (Min and Duan, 2004) shows that the contributions of single scattering to the total radiance are about 28, 50, and 95% for an atmosphere with total optical depth of 1 and single-scattering albedos of 1, 0.5, and 0.1, respectively. Figure 3.2 shows the single and multiple scattering components of radiances across the

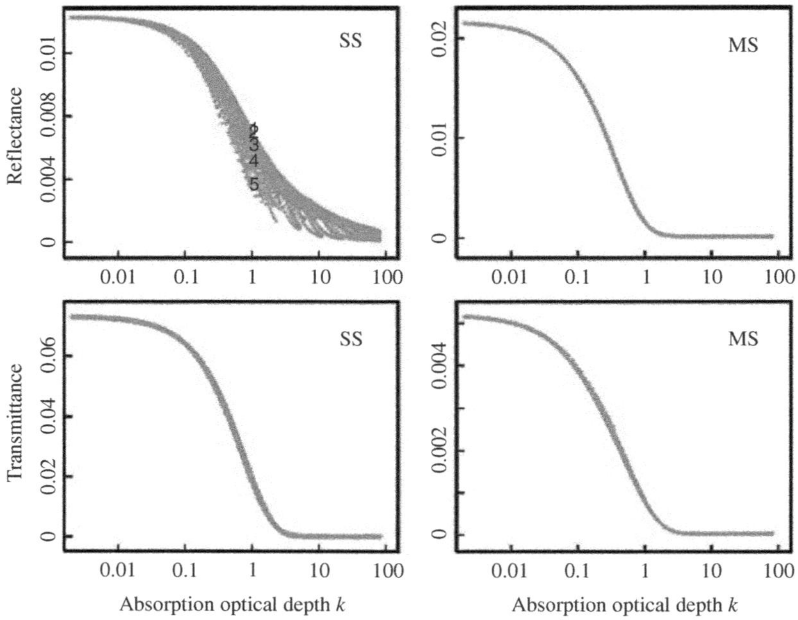

Figure 3.2 Single scattering (SS) and multiple scattering (MS) components of radiance as a function of oxygen absorption optical depths for the O_2 A-band for a clear sky with background aerosol loading. The top panels are for the reflectance at the TOA; the bottom panels are for transmittance at the surface. The solar zenith angle (SZA) is 35°, viewing zenith angle is 0°, surface albedo is 0.02, and aerosol optical depth is set to be 0.08. The numbers 1–5 in the top left plot correspond to different subranges within the O_2 A-band. *Source*: Used with permission from Duan (2005). © American Geophysical Union.

oxygenA-band as a function of oxygen absorption optical depths for the reflectance at the top of the atmosphere (TOA) and for the transmittance at the surface in a clear-sky case. Even in the case of a cirrus cloud with an optical depth of 2, as shown in Figure 3.3, the ratio of the single-scattering component to the multiple-scattering component decreases substantially with increasing absorption optical depth.

The single-scattering radiances are compared with the exact solution of the radiative transfer equation in Figures 3.4–3.6, for optical thicknesses 0.1, 0.3, and 1.0, respectively. When the optical thickness is small, tmost of the scattered radiation is due to single scattering. As the optical thickness increases, the difference between the single-scattering approximation and the exact solution increases.

3.2 Second-order Scattering Approximation

The second-order scattering approximation can be derived using the fact that the singly scattered light acts as a source for light scattered twice. In particular, Eq. (2.59) can be written in the following form:

$$\cos\vartheta\frac{dI(\vartheta,\varphi)}{d\tau}+I(\vartheta,\varphi)=B(\tau), \tag{3.25}$$

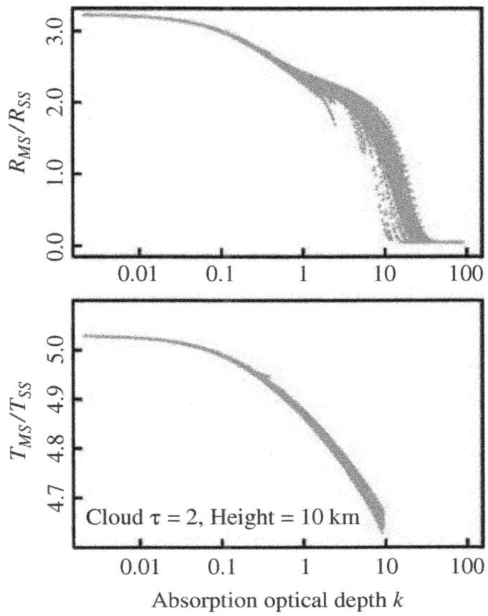

Figure 3.3 Ratio between single- and multiple-scattering components of radiance. The top panel is for the reflectance at the TOA; the bottom panel is for transmittance at the surface. The cloud is set between 9.335 and 10.05 km with an optical depth of 2, and the geometry is the same as in Figure 3.2. *Source*: Used with permission from Duan (2005). © American Geophysical Union.

where

$$B(\tau) = \frac{\omega_0}{4\pi}\int_0^{2\pi} d\varphi' \int_0^{\pi} d\vartheta' p(\Theta) I(\vartheta', \varphi') + \frac{\omega_0}{4\pi} p(\Theta_s) F_0 \exp\left(-\frac{\tau}{\cos\vartheta_0}\right) \tag{3.26}$$

is the source function. Assuming that there is no diffuse light incident on the top and bottom of a turbid layer, we derive from Eq. (3.25) for the downward and upward propagating light at its boundaries, respectively:

$$I_\downarrow(\tau, \eta, \xi, \varphi) = \int_0^{\tau} B(\tau', \eta, \xi, \varphi)\exp(-(\tau - \tau')/\eta) d\tau'/\eta, \tag{3.27}$$

$$I_\uparrow(\tau, \eta, \xi, \varphi) = \int_{\tau}^{\tau_0} B(\tau', \eta, \xi, \varphi)\exp((\tau - \tau')/\eta) d\tau'/\eta, \tag{3.28}$$

where $\eta = |\cos\vartheta|$, $\xi = \cos\vartheta_0$, τ_0 is the optical thickness of the scattering layer. Assuming that I in Eq. (3.26) can be presented in the single-scattering approximation derived in the previous section, one can find intensity in the secondary scattering approximation using Eqs. (3.27) and (3.28). In a similar way, all orders of scattering can be accounted for. It follows after N scattering events:

$$I_\downarrow(\tau, \eta, \xi, \varphi) = \sum_{n=0}^{N} \omega_0^n I_{\downarrow,n}, \quad I_\uparrow(\tau, \eta, \xi, \varphi) = \sum_{n=0}^{N} \omega_0^n I_{\uparrow,n} \tag{3.29}$$

or for transmission and reflection functions defined earlier:

$$T(\tau, \eta, \xi, \varphi) = \sum_{n=0}^{N} \omega_0^n \tilde{T}_n, \quad R(\tau, \eta, \xi, \varphi) = \sum_{n=0}^{N} \omega_0^n \tilde{R}_n. \tag{3.30}$$

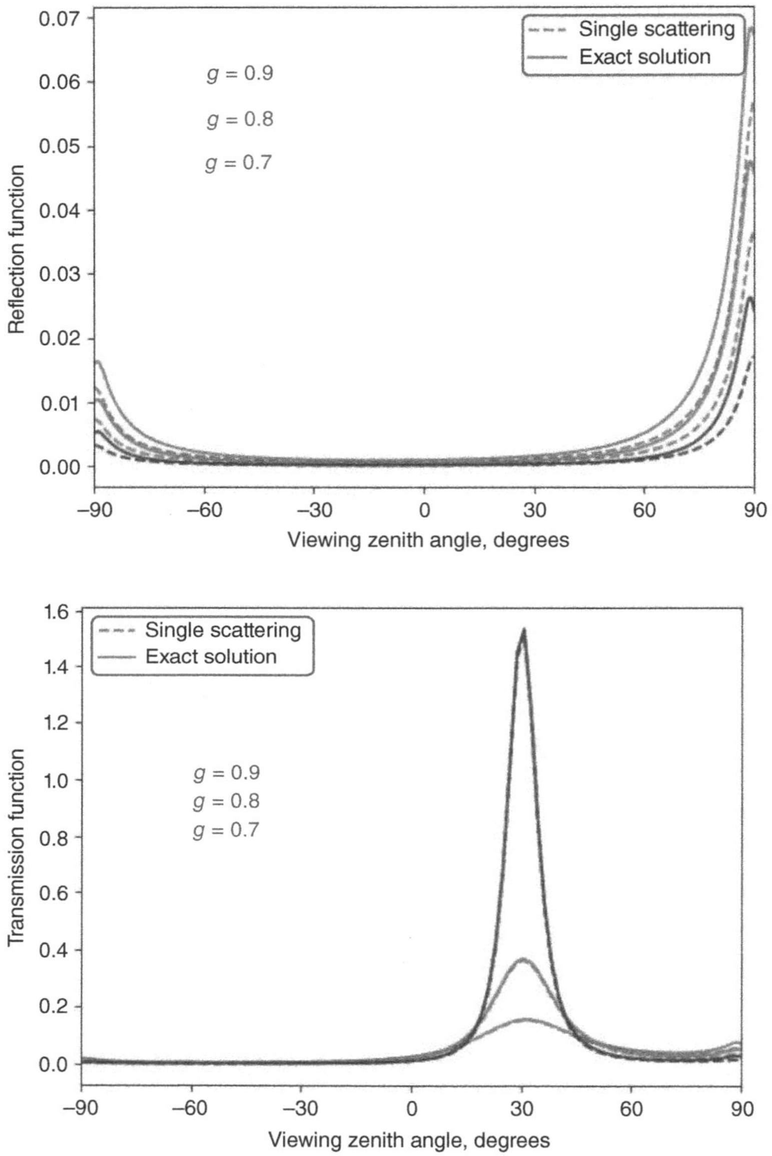

Figure 3.4 Reflected (top) and transmitted (bottom) radiances computed by the single-scattering approximation and the exact solution for different values of the asymmetry parameter. The optical thickness of the layer is 0.1, the single-scattering albedo is 0.99, and the angle of incidence is 30°.

The functions $\tilde{T}_1, \tilde{R}_1$ are calculated on the basis of the single-scattering approximation (Eqs. (3.18)–(3.19)). The expressions for $\tilde{T}_2, \tilde{R}_2$ are given by Hovenier et al. (2004). Hansen and Travis (1974) have proposed the use of invariance principles for the determination of the

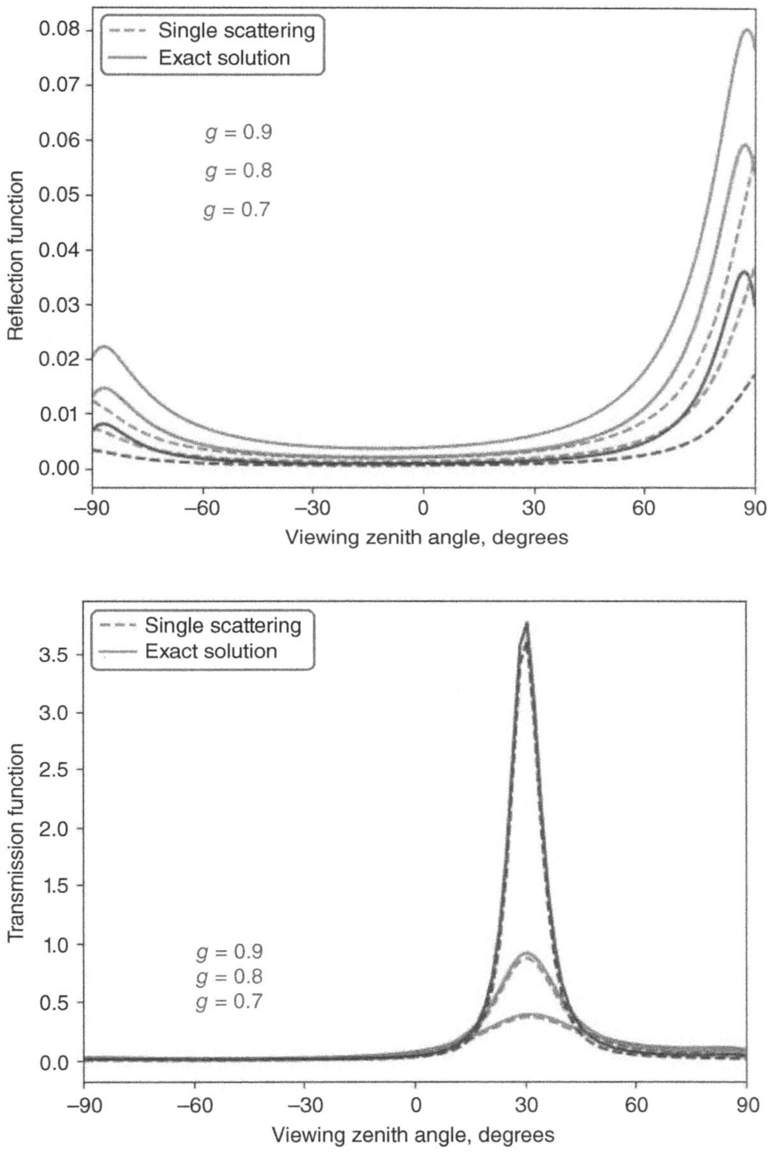

Figure 3.5 Same as Figure 3.4, but for an optical thickness of 0.3.

couplet $(\tilde{T}_2, \tilde{R}_2)$. The final equations have the following forms for vertically homogeneous turbid layers:

$$\tilde{T}_2(\tau, \eta, \xi, \varphi) = \frac{\xi\mu}{4\pi(\xi + \mu)} \int_0^1 d\eta' \int_0^{2\pi} \Phi_T(\eta', \varphi') d\varphi', \tag{3.31}$$

$$\tilde{R}_2(\tau, \eta, \xi, \varphi) = \frac{\xi\mu}{4\pi(\xi - \mu)} \int_0^1 d\eta' \int_0^{2\pi} \Phi_R(\eta', \varphi') d\varphi', \tag{3.32}$$

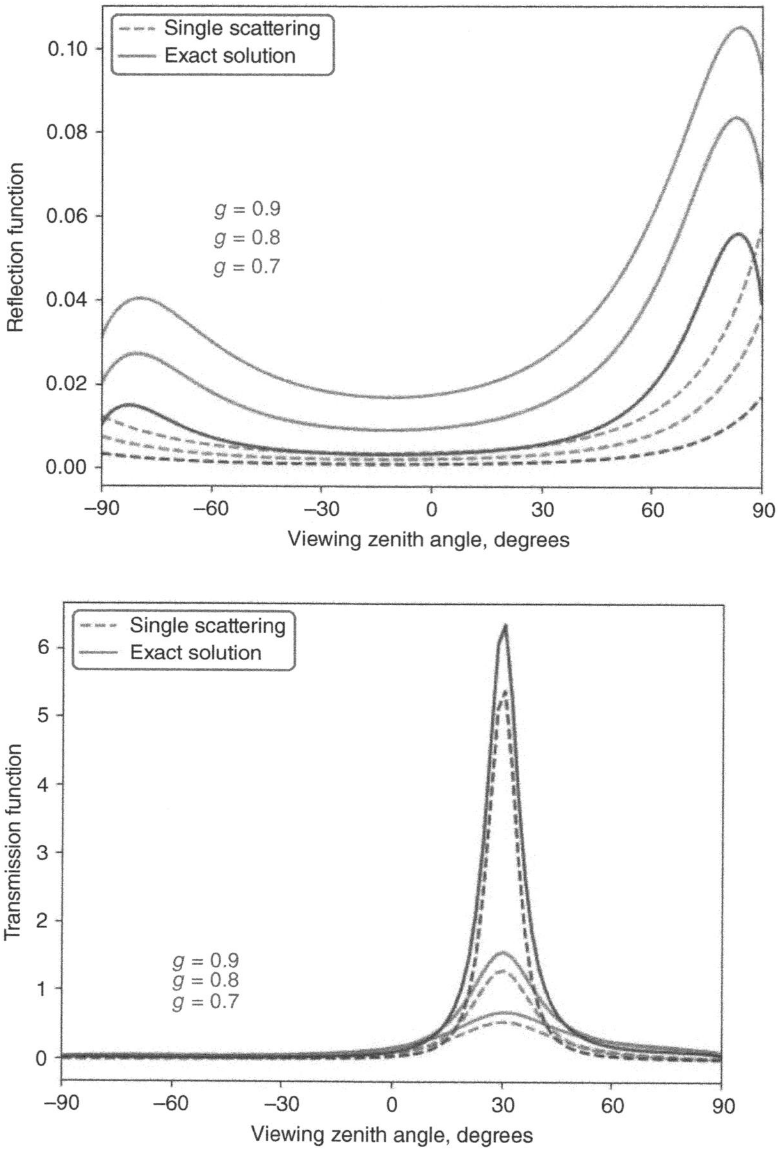

Figure 3.6 Same as Figure 3.4, but for an optical thickness of 1.0.

where

$$\begin{aligned}\Phi_T(\eta',\varphi') &= \eta^{-1}p_t(\eta,\eta',\varphi'-\varphi)\tilde{T}_1(\eta',\xi,\varphi'-\varphi_0)\\ &-\xi^{-1}p_t(\xi,\eta',\varphi'-\varphi_0)\tilde{T}_1(\eta,\eta',\varphi'-\varphi)-\Xi_T(\eta',\varphi'),\end{aligned} \tag{3.33}$$

$$\begin{aligned}\Phi_R(\eta',\varphi') &= \eta^{-1}p_t(\eta,\eta',\varphi'-\varphi)\tilde{R}_1(\eta',\xi,\varphi'-\varphi_0)\\ &+\xi^{-1}p_t(\xi,\eta',\varphi'-\varphi_0)\tilde{R}_1(\eta,\eta',\varphi'-\varphi)-\Xi_R(\eta',\varphi'),\end{aligned} \tag{3.34}$$

$$\begin{aligned}\Xi_T(\eta',\varphi') &= \eta^{-1}\exp(-\tau/\eta)p_r(\eta,\eta',\varphi'-\varphi)\tilde{R}_1(\xi,\eta',\varphi'-\varphi)\\&-\xi^{-1}\exp(-\tau/\xi)p_r(\xi,\eta',\varphi'-\varphi_0)\tilde{R}_1(\eta,\eta',\varphi-\varphi'),\end{aligned} \tag{3.35}$$

$$\begin{aligned}\Xi_R(\eta',\varphi') &= \eta^{-1}\exp(-\tau/\eta)p_r(\eta,\eta',\varphi'-\varphi)\tilde{T}_1(\xi,\eta',\varphi'-\varphi_0)\\&+\xi^{-1}\exp(-\tau/\xi)p_r(\xi,\eta',\varphi'-\varphi_0)\tilde{T}_1(\eta,\eta',\varphi'-\varphi),\end{aligned} \tag{3.36}$$

$$\tilde{T}_1 = \frac{p_t}{4(\xi-\eta)}\left\{\exp\left[-\frac{\tau_0}{\xi}\right]-\exp\left[-\frac{\tau_0}{\eta}\right]\right\}, \tag{3.37}$$

$$\tilde{R}_1 = \frac{p_r}{4(\xi+\eta)}\left\{1-\exp\left[\left(\frac{1}{\xi}+\frac{1}{\eta}\right)\tau_0\right]\right\}. \tag{3.38}$$

Let us note that Eqs. (3.37)–(3.38) have been derived earlier (see Eqs. (3.18)–(3.19)). Also, it follows at $\xi = \eta$:

$$\tilde{T}_1 = \frac{\tau_0 p(\Theta)}{4\xi^2}\exp\left[-\frac{\tau_0}{\xi}\right] \tag{3.39}$$

The values of p_t and p_r are the single-scattering phase functions evaluated at the cosines of the scattering angle $\Theta_t(\xi,\eta,\varphi,\varphi_0)$ and $\Theta_r(\xi,\eta,\varphi,\varphi_0)$ characterized by the following equations:

$$\cos\Theta_t = \xi\eta + \sqrt{(1-\xi^2)(1-\eta^2)}\cos(\varphi-\varphi_0) \tag{3.40}$$

and

$$\cos\Theta_r = -\xi\eta + \sqrt{(1-\xi^2)(1-\eta^2)}\cos(\varphi-\varphi_0), \tag{3.41}$$

respectively. Relevant expressions for the Fourier components of the reflection and transmission functions of various orders of scattering were obtained by Kawabata and Ueno (1988). The aforementioned expressions provide useful approximations for optically thin media. They are also useful to compute reflection and transmission functions for the initial thin layer in adding-doubling computations. Using these expressions, Mei et al. (2020) developed a fast model for retrieving aerosol properties.

3.3 Small-angle Approximation

According to Mie theory (Mie, 1908), in the case of light scattering by particles that are much larger than the wavelength of incident light, the single-scattering phase function is forward-peaked, meaning that most of the photons are scattered in the directions close to the forward one – that is, small-angle scattering. Figure 3.7 shows an example of the phase function computed for a water cloud, for which 98.7% photons are scattered within the angular range of 0–3°.

The small-angle approximation can be applied for computing the transmitted light intensity in the directions close to that of the incident light beam. In practice, the accuracy of this approach increases as the single-scattering phase function becomes more forward-peaked.

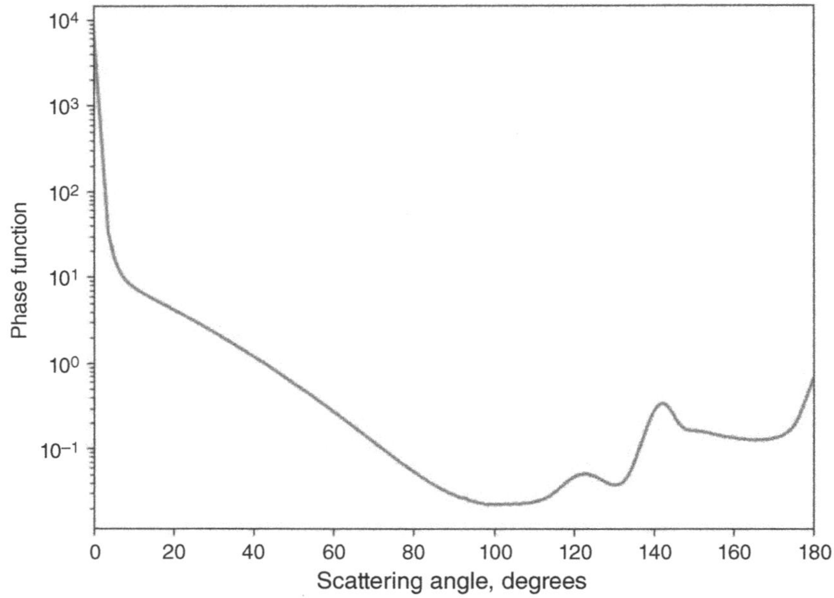

Figure 3.7 Single-scattering phase function for a cloud consisting of an ensemble of spherical water droplets. The particle radius follows the modified gamma distribution (Hess et al., 1998).

One way to apply the small-angle approximation is to substitute μ with μ_0 on the left-hand side of the RTE. Using this idea for Eq. (2.75), we obtain the RTE for azimuthal expansion coefficients I_m of the radiance field in the small-angle approximation:

$$\mu_0 \frac{\mathrm{dI}_m}{\mathrm{d}\tau}(\tau,\mu) = -\mathrm{I}_m(\tau,\mu) + \frac{\omega_0}{2}\int_{-1}^{1} \mathrm{I}_m(\tau,\mu')\, p_m(\mu,\mu')\, d\mu'. \tag{3.42}$$

Note that here I_m depends on the spatial coordinate. Equation (3.42) can be solved analytically, and several interesting results can be obtained, e.g. including studies of optical transfer and coherence functions (Dolin, 1964). We expand the radiance I_m into Legendre series:

$$\mathrm{I}_m(\tau,\mu) = \sum_{n=m}^{\infty} (2n+1)\frac{(n-m)!}{(n+m)!}\mathrm{I}_{mn}(\tau)\, P_n^m(\mu)P_n^m(\mu'). \tag{3.43}$$

The same procedure is applied for the single-scattering phase function:

$$p_m(\mu,\mu') = \sum_{n=m}^{\infty} (2n+1)\frac{(n-m)!}{(n+m)!}\chi_n P_n^m(\mu)P_n^m(\mu'). \tag{3.44}$$

Note that on the left-hand side of Eq. (3.42) we have $\mu_0 \mathrm{dI}/\mathrm{d}\tau$ instead of $\mu dI/\mathrm{d}\tau$, and thus, all terms in Eq. (3.42) have the same expansion kernels.

Substituting Eq. (3.43) to Eq. (3.42) and applying the orthogonality property of the Legendre functions (A.4), we have:

$$\mu_0 \frac{\mathrm{dI}_{nm}(\tau)}{\mathrm{d}\tau} = -\mathrm{I}_{nm}(\tau) + \omega\chi_n \mathrm{I}_{nm}(\tau) \tag{3.45}$$

The solution is

$$I_{nm}(\tau) = C_{nm} e^{-\frac{1}{\mu_0}(1-\omega\chi_n)\tau} \tag{3.46}$$

where C_{nm} are integration constants that are to be defined from the boundary conditions. For the following boundary conditions at the top of the medium:

$$\mathrm{I}(\vartheta, \varphi) = F\delta(\mu - \mu_0)\,\delta(\varphi - \varphi_0) \tag{3.47}$$

the final solution reads as follows:

$$\begin{aligned} I(\tau, \vartheta, \varphi) = \frac{F}{4\pi} \sum_{m=0}^{\infty} \sum_{n=m}^{\infty} \Bigg\{ & (2 - \delta_{m0})(2n+1) \frac{(n-m)!}{(n+m)!} \\ & \times e^{-\frac{1}{\mu_0}(1-\omega\chi_n)\tau} P_n^m(\mu) P_n^m(\mu_0) \cos(m\varphi) \Bigg\}. \end{aligned} \tag{3.48}$$

The small-angle solution gives a Dirac delta function at the upper boundary and a peaked function inside the scattering medium. The amplitude of the peaked function decreases with increasing optical depth, while its width increases with increasing optical depth.

We note that, similar to the direct solar beam, the small-angle solution solves the boundary-value problem for a semi-infinite medium since no reflection from the lower boundary is allowed. Note that Eq. (3.48) contains a singularity term due to the contribution of the direct solar beam (see Eq. (2.58)). The solution for the diffuse component reads as follows:

$$\begin{aligned} I(\tau, \vartheta, \varphi) = \frac{F}{4\pi} \sum_{m=0}^{\infty} \sum_{n=m}^{\infty} & (2 - \delta_{m0})(2n+1) \frac{(n-m)!}{(n+m)!} \\ & \times \left[e^{-\frac{1}{\mu_0}(1-\omega\chi_n)\tau} - e^{-\frac{1}{\mu_0}\tau} \right] P_n^m(\mu) P_n^m(\mu_0) \cos(m\varphi). \end{aligned} \tag{3.49}$$

For the case of normal incidence, we have $\mu_0 = 1$, and all azimuthal harmonics $m > 0$ vanish. Consequently, Eq. (3.49) is simplified:

$$I_0(\tau, \vartheta) = \frac{F}{4\pi} \sum_{n=0}^{\infty} (2n+1) \left[e^{-\frac{1}{\mu_0}(1-\omega\chi_n)\tau} - e^{-\frac{1}{\mu_0}\tau} \right] P_n(\mu). \tag{3.50}$$

Equation (3.50) can be rewritten in the integral form (Kokhanovsky, 1997). We use the formula for expansion coefficient χ_n:

$$\chi_n = \frac{1}{2} \int_{-1}^{1} P_n(\mu)\, p(\mu)\, d\mu. \tag{3.51}$$

Then, substituting Eq. (3.51) into Eq. (3.50) and making use of the asymptotic formula (Szegö, 1992)

$$\lim_{\theta \to 0} P_n(\mu) = J_0\left(\theta\left(n + \frac{1}{2}\right)\right), \tag{3.52}$$

where $\mu = \cos\theta$, we obtain:

$$I(\tau,\vartheta) = \frac{F}{4\pi}\sum_{n=0}^{\infty} 2\left(n+\frac{1}{2}\right)\left[e^{-\frac{1}{\mu_0}\left(1-\omega\frac{1}{2}\int_{-1}^{1} J_0\left(\theta\left(n+\frac{1}{2}\right)\right)p(\mu)\,d\mu\right)\tau} - e^{-\frac{1}{\mu_0}\tau}\right] \times J_0\left(\theta\left(n+\frac{1}{2}\right)\right). \tag{3.53}$$

Introducing a new function:

$$f\left(n+\frac{1}{2}\right) = \left(n+\frac{1}{2}\right)\left[e^{-\frac{1}{\mu_0}\left(1-\omega\frac{1}{2}\int_{-1}^{1} J_0\left(\theta\left(n+\frac{1}{2}\right)\right)p(\mu)\,d\mu\right)\tau} - e^{-\frac{1}{\mu_0}\tau}\right] \times J_0\left(\theta\left(n+\frac{1}{2}\right)\right), \tag{3.54}$$

Eq. (3.53) can be rewritten as follows:

$$I(\tau,\mu) = \frac{F}{2\pi}\sum_{n=0}^{\infty} f\left(n+\frac{1}{2}\right). \tag{3.55}$$

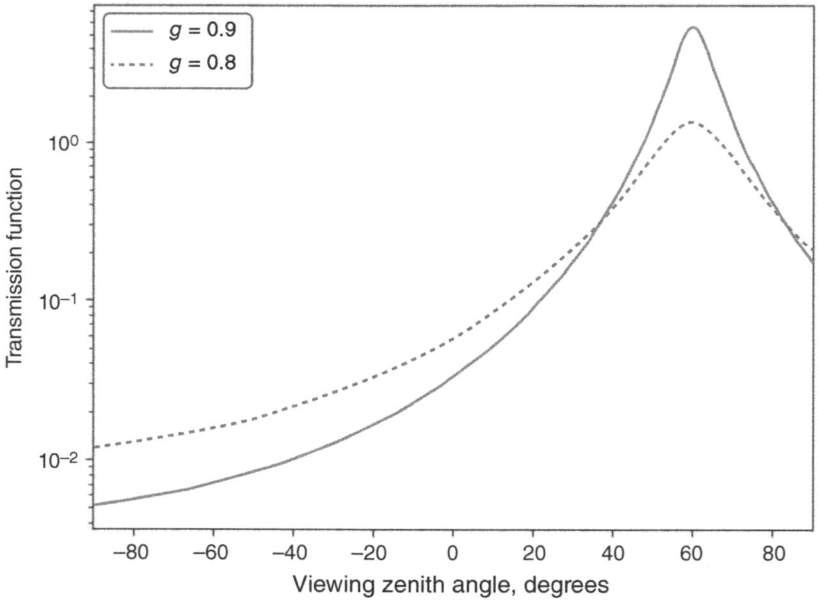

Figure 3.8 Transmitted radiances computed using the small-angle approximation. The computations are performed for the Henyey–Greenstein phase function with asymmetry parameter values of 0.9 and 0.8. The single-scattering albedo is 0.99. The angle of incidence is 60°. The optical thickness is 1.0.

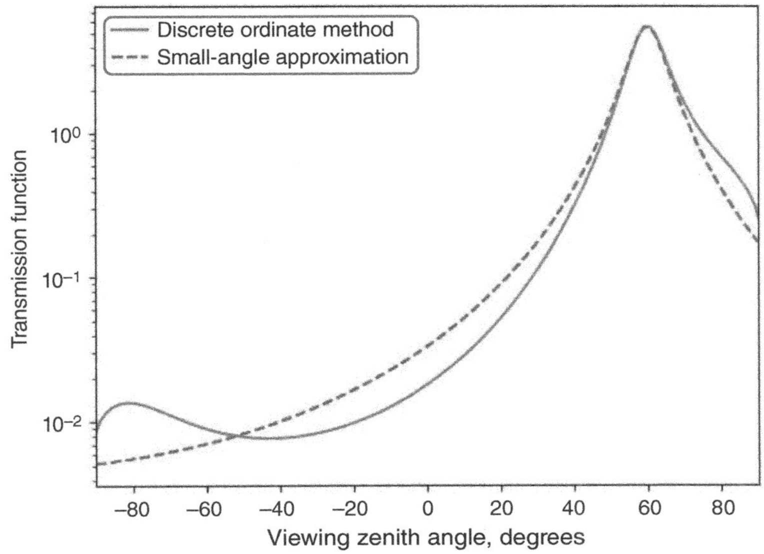

Figure 3.9 Comparison between solutions derived using the discrete ordinates method and the small-angle approximation. The computations are performed for the Henyey–Greenstein phase function with asymmetry parameter 0.9. The single-scattering albedo is 0.99. The angle of incidence is 60°. The optical thickness is 1.0.

Then, approximating the sum by an integral, i.e.

$$\sum_{n=0}^{\infty} f\left(n + \frac{1}{2}\right) = \int_0^{\infty} f(\xi)\, d\xi, \tag{3.56}$$

the small-angle solution can be represented in the form:

$$I(\tau, \vartheta) = \frac{F}{2\pi} \int_0^{\infty} f(\xi)\, d\xi. \tag{3.57}$$

The advantage of the integral solution (3.57) is that it can be analytically integrated for special types of phase functions (Zege and Kokhanovsky, 1994).

Figures 3.8 and 3.9 illustrate the radiances computed with the small-angle approximation. As we can see, it agrees well with the discrete ordinates solution for angles close to the incident angle.

4

Semi-infinite Media

4.1 Milne Problem

The Milne problem, named after British mathematician Edward Arthur Milne (Milne, 1921), is a classic problem in radiative transfer theory. It serves as a fundamental example to understand how radiation propagates through a medium and how it is modified by scattering and absorption processes. In the Milne problem, the scenario involves a semi-infinite, homogeneous medium that is illuminated by an external radiation source. The medium is assumed to be both absorbing and scattering.

Mathematically, the Milne problem can be expressed as a boundary value problem for the radiative transfer equation (RTE) with appropriate boundary conditions. Typically, these boundary conditions specify the incident radiation flux at the surface of the medium and may also include conditions at infinite depth (e.g. vanishing radiation intensity).

Let us consider the RTE for azimuthally averaged radiance for a nonabsorbing medium:

$$\mu\frac{dI(\tau,\mu)}{d\tau} = -I(\tau,\mu) + \frac{1}{2}\int_{-1}^{1} p(\mu,\mu')I(\tau,\mu')d\mu'. \tag{4.1}$$

In the case of isotropic scattering, we have:

$$p(\mu,\mu') = 1 \tag{4.2}$$

and Eq. (4.1) may be written as follows:

$$\mu\frac{dI(\tau,\mu)}{d\tau} = -I(\tau,\mu) + J(\tau,\mu), \tag{4.3}$$

where

$$J(\tau,\mu) = \frac{1}{2}\int_{-1}^{1} I(\tau,\mu)d\mu. \tag{4.4}$$

Analytical Methods in Radiative Transfer, First Edition. Alexander Kokhanovsky, Vijay Natraj and Dmitry Efremenko.

Note that $J(\tau,\mu)$ can be associated with the mean radiance (see Eq. (2.80)). If the medium is not illuminated by any external radiation field, the formal solution may be written for the upwelling radiation ($\mu < 0$):

$$I(\tau,\mu) = \int_{\tau}^{\infty} e^{-(t-\tau)/|\mu|} J(t) \frac{dt}{|\mu|} \tag{4.5}$$

and downwelling radiation ($\mu > 0$):

$$I(\tau,\mu) = \int_{0}^{\tau} e^{-(\tau-t)/\mu} J(t) \frac{dt}{\mu}. \tag{4.6}$$

Then let us consider the radiance momentum of the n-th order (see Eq. (2.79)) and take the integrals over the upper and lower hemispheres separately, namely

$$2M_n(\tau) = \int_{-1}^{1} I(\tau,\mu)\mu^n d\mu = \int_{-1}^{0} I(\tau,\mu)\mu^n d\mu + \int_{0}^{1} I(\tau,\mu)\mu^n d\mu. \tag{4.7}$$

Substituting Eqs. (4.5) and (4.6) into Eq. (4.7), we get

$$\begin{aligned} 2M_n(\tau) = (-1)^n \int_{\tau}^{\infty} d\tau J(\tau) \int_{-1}^{0} e^{-(t-\tau)/|\mu|} J(t) \frac{dt}{|\mu|} \\ + \int_{0}^{\tau} d\tau J(\tau) \int_{0}^{1} e^{-(\tau-t)/\mu} J(t) \frac{dt}{\mu}. \end{aligned} \tag{4.8}$$

Then, using substitution

$$x = \frac{1}{\mu}, \tag{4.9}$$

Eq. (4.10) becomes

$$\begin{aligned} 2M_n(\tau) = (-1)^n \int_{\tau}^{\infty} d\tau J(\tau) \int_{1}^{\infty} e^{-(t-\tau)x} \frac{dt}{x^{n+1}} \\ + \int_{0}^{\tau} d\tau J(\tau) \int_{0}^{1} e^{-(\tau-t)x} \frac{dt}{x^{n+1}}. \end{aligned} \tag{4.10}$$

Let us note that the function

$$E_n(q) = \int_{1}^{\infty} \frac{e^{-xq}}{x^n} dx \tag{4.11}$$

is referred to as the exponential integral, and Eq. (4.10) may be written as follows:

$$M_n(\tau) = (-1)^n \int_{\tau}^{\infty} d\tau J(\tau) E_{n+1}(t-\tau) + \int_{0}^{\tau} d\tau J(\tau) E_{n+1}(\tau - t). \tag{4.12}$$

Recalling that $M_0(\tau) = J(\tau)$, for $n = 0$, we have

$$J(\tau) = \frac{1}{2}\int_0^\infty J(t)E_1(|\tau - t|)dt. \tag{4.13}$$

This is the Schwarzschild–Milne equation. A function that multiplies the unknown in the integrand of an integral equation is referred to as the kernel of the integral equation. Thus, the exponential integral in Eq. (4.13) is the kernel of the integral equation for the source function. The Schwarzschild–Milne equation is fundamental in the field of astrophysics, particularly in modeling stellar atmospheres and interpreting astronomical observations. It allows astrophysicists to predict how stars of different compositions, temperatures, and pressures emit light. The equation is also used to estimate stellar ages, compositions, and distances by comparing predicted spectra with observed spectra. Also, note that if the source J is found from Eq. (4.13), then the radiance can be computed using Eqs. (4.5) and (4.6).

4.2 Light in Deep Layers of Semi-infinite Turbid Media

Let us consider an important radiative transfer regime occurring in deep layers of semi-infinite light-absorbing media. It is known from experimental measurements that light intensity attenuates exponentially with the angular pattern $i(\eta)$, which does not depend on the optical depth in this deep layer regime. Also, the dependence on the azimuthal angle disappears. It follows that:

$$I(\eta) = ci(\eta)\exp(-k\tau), \tag{4.14}$$

where c is a constant. The task of the theory is to derive the angular distribution $i(\eta)$ and also the diffusion exponent k as the functions of local optical parameters of a turbid medium under consideration. Equation (2.59) takes the following form in deep layers, where we have neglected the single light scattering contribution and also the azimuthal dependence of light intensity characteristic of an asymptotic regime discussed in this section:

$$\cos\vartheta\frac{dI(\vartheta)}{d\tau} - \frac{\omega_0}{2}\int_{-1}^{1} d\vartheta' p(\cos\vartheta', \cos\vartheta)I(\cos\vartheta') + I(\cos\vartheta) = 0 \tag{4.15}$$

It follows after the substitution of Eq. (4.14) into Eq. (4.15):

$$(1 - k\eta)i(\eta) = \frac{\omega_0}{2}\int_{-1}^{1} p(\eta, \eta')i(\eta')\,d\eta', \tag{4.16}$$

where

$$p(\eta, \eta') = \frac{1}{2\pi}\int_0^{2\pi} p(\cos\Theta)d\varphi \tag{4.17}$$

is the azimuthally averaged phase function. Let us assume that the turbid medium is characterized by isotropic scattering and, therefore, $p(\eta, \eta') = 1$. Also, we shall use the normalization condition:

$$\frac{\omega_0}{2} \int_{-1}^{1} i(\eta')\, d\eta' = 1. \tag{4.18}$$

Then we derive from Eq. (4.16):

$$i(\eta) = \frac{1}{1 - k\eta}. \tag{4.19}$$

This function is presented in Figure 4.1 for several values of k. It follows that in the case of nonabsorbing isotropically scattering media, the light intensity pattern has a spherical shape and $i(\eta) = 1$. For absorbing media, the light intensity monotonically decreases from the nadir direction toward oblique observation directions. The diffusion coefficient k can be derived from measurements of light intensity at two values of the observation angle. Let us substitute Eq. (4.19) into Eq. (4.18). Then one can derive:

$$\omega_0^{-1} = \frac{1}{2k} \ln\left(\frac{1+k}{1-k}\right). \tag{4.20}$$

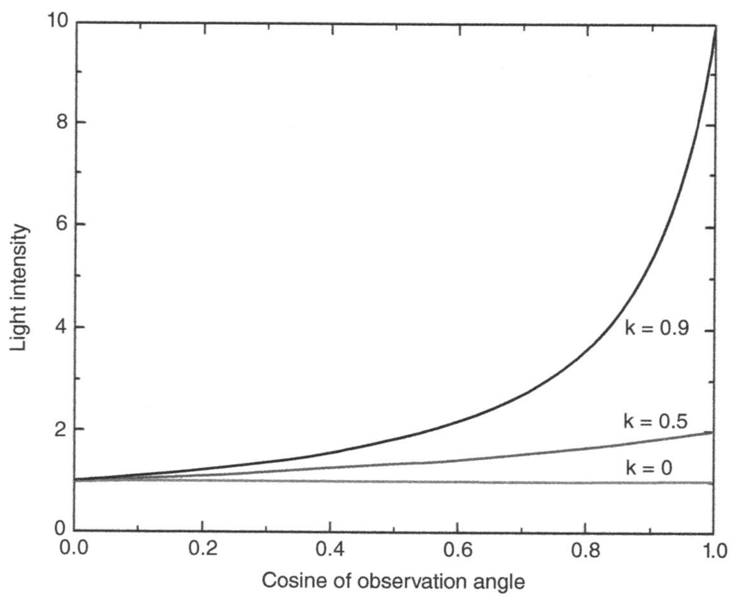

Figure 4.1 Angular dependence of light intensity in deep layers of turbid media with isotropic scattering.

This equation can be used to derive the dependence of the diffusion exponent on the single-scattering albedo (SSA) ω_0 in the case of isotropic scattering. It follows that at $k\rightarrow 0$:

$$\frac{1}{2k}\ln\left(\frac{1+k}{1-k}\right)=\frac{1}{1-k^2/3} \tag{4.21}$$

and, therefore:

$$k=\sqrt{3(1-\omega_0)}. \tag{4.22}$$

This expression is valid for small values of the probability of photon absorption $\beta = 1-\omega_0$ for an isotropic scattering media. It follows from Eq. (4.19) for weakly absorbing media:

$$i(\eta)=1+k\cos\vartheta. \tag{4.23}$$

Most turbid media scatter light predominantly in the forward scattering direction and Eqs. (4.19)–(4.22) must be modified.

Let us consider the following phase function:

$$p(\Theta)=1+3g\cos\Theta, \tag{4.24}$$

where

$$g=\frac{1}{2}\int_0^\pi p(\Theta)\sin\Theta\cos\Theta d\Theta \tag{4.25}$$

is the asymmetry parameter. Then one can derive, assuming that I is given by Eq. (4.14):

$$(1-k\mu)i(\mu)=\frac{\omega_0}{2}\int_{-1}^{1}(1+x_1\mu\mu')i(\mu')\,d\mu', \tag{4.26}$$

where we accounted for the fact that the azimuthally averaged cosine of the scattering angle can be substituted by the product $\mu\mu'$. Therefore, we derive:

$$i(\mu)=\frac{b(\mu)}{1-k\mu}, \tag{4.27}$$

where

$$b(\mu)=\frac{\omega_0}{2}\int_{-1}^{1}(1+x_1\mu\mu')i(\mu')\,d\mu' \tag{4.28}$$

or

$$b(\mu)=\frac{\omega_0}{2}\int_{-1}^{1}\frac{(1+x_1\mu\mu')b(\mu')}{1-k\mu'}d\mu'. \tag{4.29}$$

It is clear from Eq. (4.29) that $b(\mu)$ can be presented in the following form:

$$b(\mu)=b_0+b_1\mu, \tag{4.30}$$

where

$$b_0 = \frac{\omega_0}{2}\int_{-1}^{1}\frac{b(\mu')d\mu'}{1-k\mu'}, \tag{4.31}$$

$$b_1 = \frac{\omega_0 x_1}{2}\int_{-1}^{1}\frac{\mu' b(\mu')}{1-k\mu'}d\mu'. \tag{4.32}$$

It follows from Eq. (4.30) and (4.31):

$$b_0 + b_1\mu = b_0 j_0 + b_1 j_1 + b_0 x_1 j_1 \mu + b_1 x_1 j_2 \mu \tag{4.33}$$

or

$$b_0 = b_0 j_0 + b_1 j_1, \tag{4.34}$$

$$b_1 = b_0 x_1 j_1 + b_1 x_1 j_2, \tag{4.35}$$

where

$$j_n = \frac{\omega_0}{2}\int_{-1}^{1}\frac{(\mu')^n d\mu'}{1-k\mu'}, \tag{4.36}$$

or after integration,

$$j_0 = \omega_0 s, \tag{4.37}$$

$$j_1 = \frac{\omega_0(s-1)}{k}, \tag{4.38}$$

$$j_2 = \frac{j_1}{k}. \tag{4.39}$$

Therefore, we have the following equation for the determination of both b_0 and b_1:

$$\hat{a}\vec{b} = 0, \tag{4.40}$$

where the elements of the matrix $\hat{a}$ can be represented in the following way:

$$a_{11} = \frac{1}{\omega_0} - s, \tag{4.41}$$

$$a_{12} = a_{21} = \frac{1-s}{k}, \tag{4.42}$$

$$a_{22} = \frac{1}{\omega_0 x_1} - \frac{s-1}{k^2}. \tag{4.43}$$

This equation has a solution if the determinant

$$D = a_{11}a_{22} - a_{12}a_{21} \tag{4.44}$$

is equal to zero. Therefore, we have

$$\omega_0 \left[1 + x_1 \frac{1-\omega_0}{k^2}\right] s - \omega_0 x_1 \frac{1-\omega_0}{k^2} = 1. \tag{4.45}$$

Equation (4.45) can be used to determine the diffusion exponent for any set of values of the doublet (x_1, ω_0). Let us assume that the medium under consideration is weakly absorbing. Then the probability of photon absorption $\beta = 1 - \omega_0 \to 0$, while $s \to 1 + \frac{k^2}{3}$ and, therefore, one can derive:

$$k = \sqrt{3\beta(1-g)}, \tag{4.46}$$

which reduces to the value derived for the case of isotropic scattering ($g = 0$), as expected. Importantly, although this expression for the diffusion exponent in a weakly absorbing medium has been derived for the case of a specific phase function, it can be used for any arbitrary phase function under the assumption that $\beta \to 0$. It follows for the function $i(\mu)$:

$$i(\mu) = b_0 \frac{1+\xi\mu}{1-k\mu}, \tag{4.47}$$

where $\xi = b_1/b_0$. This function can be multiplied by any constant and still be a solution of Eq. (4.26). Let us assume that $b_0 = 1$ and $\xi = b_1$. Then it follows from Eq. (4.34):

$$\xi = x_1 j_1 (1 + \frac{\xi}{k}) \tag{4.48}$$

or

$$\xi = \frac{\Pi}{1 - \Pi/k}, \tag{4.49}$$

where $\Pi = (s-1)\omega_0 x_1 k^{-1}$. Let us assume that $\beta \to 0$. Then one can derive:

$$\xi = \frac{gk\omega_0}{1 - g\omega_0} \tag{4.50}$$

and

$$i(\mu) = 1 + 3s\mu, \tag{4.51}$$

where

$$s = \frac{\beta}{k} \tag{4.52}$$

is the similarity parameter. One can see that the ratio of the probability of photon absorption to the diffusion exponent is a key parameter for radiative transfer in snow. Also, we can write:

$$s = \sqrt{\frac{\beta}{3(1-g)}} \equiv \frac{k}{3(1-g)}. \tag{4.53}$$

The parameters (k,s) can be used to derive the intensity of light in deep layers of weakly absorbing media with the phase function given by Eq. (4.24). It follows that:

$$i(\mu) = (1 + 3s\mu)\exp(-k\tau) \tag{4.54}$$

as $\beta \to 0$. This equation can also be written in the following form:

$$I(\mu) = (1 + 3s\mu)\exp(-\gamma h), \tag{4.55}$$

where

$$\gamma = kk_{\text{ext}} \tag{4.56}$$

is the so-called asymptotic flux attenuation coefficient and h is the geometrical depth. Using the e-folding depth $h_e = 1/\gamma$, one can derive from Eq. (4.55):

$$i(\mu) = \left(1 + \frac{k\mu}{1-g}\right)\exp(-h/h_e). \tag{4.57}$$

Although Eqs. (4.46) and (4.57) have been derived for a specific phase function, they remain valid for any arbitrary phase function as $\omega_0 > 1$.

4.3 Light Reflection from Semi-infinite Media

4.3.1 Invariant Imbedding

Invariance principles in radiative transfer were developed by Ambartsumian (1944b) and utilized by Chandrasekhar (1950). In particular, the invariance principles take into account the fact that the outgoing radiation from a medium of thickness d, when added to a medium of thickness d', should produce the same effect as a single medium of thickness $d + d'$, assuming no change in the optical properties of the medium (Nikoghossian, 2009).

To demonstrate this, consider an infinitely thin layer placed atop a semi-infinite medium. We analyze four key processes within the layer:

1. Single scattering within the thin layer.
2. Single scattering that leads to penetration into the semi-infinite medium, followed by reflection from it.
3. Reflection from the semi-infinite medium, followed by single scattering within the thin layer.
4. Reflection from the semi-infinite medium, followed by single scattering within the thin layer, and a subsequent reflection.

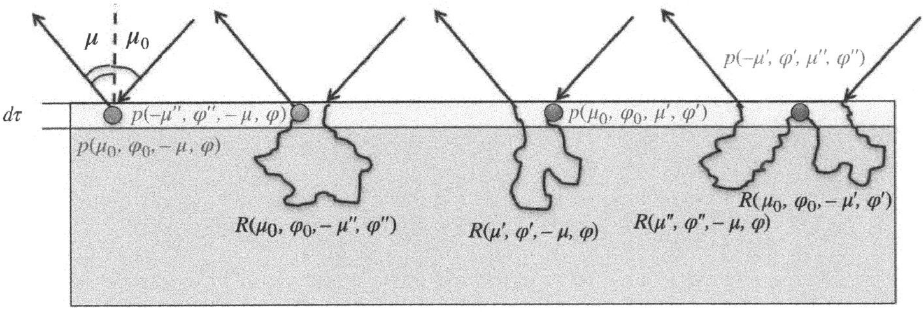

Figure 4.2 Illustration of the invariant imbedding method: four processes occurring in an infinitely thin layer that is added to a semi-infinite medium.

These processes are shown in Figure 4.2. Since the reflection function should not change, we obtain the following integral equation for the reflection function of a semi-infinite turbid layer:

$$\begin{aligned}
&\left(\frac{1}{\mu}+\frac{1}{\mu_0}\right)\hat{R}(\mu_0,\varphi_0,-\mu,\varphi)=\omega p(\mu_0,\varphi_0,-\mu,\varphi)\\
&+\frac{\omega}{4\pi}\int_0^{2\pi}\int_0^1 \hat{R}(\mu_0,\varphi_0,-\mu'',\varphi'')p(-\mu'',\varphi'',-\mu,\varphi)\frac{d\mu''}{\mu''}d\varphi''\\
&+\frac{\omega}{4\pi}\int_0^{2\pi}\int_0^1 p(\mu_0,\varphi_0,\mu',\varphi')\hat{R}(\mu',\varphi',-\mu,\varphi)\frac{d\mu'}{\mu'}d\varphi'\\
&+\frac{\omega}{16\pi^2}\int_0^{2\pi}\int_0^1\int_0^{2\pi}\int_0^1 \hat{R}(\mu_0,\varphi_0,-\mu',\varphi')p(-\mu',\varphi',\mu'',\varphi'')\\
&\times\hat{R}(\mu'',\varphi'',-\mu,\varphi_0)\frac{d\mu'}{\mu'}d\varphi'\frac{d\mu''}{\mu''}d\varphi'',
\end{aligned} \tag{4.58}$$

where

$$\hat{R}=4R\mu\mu_0. \tag{4.59}$$

The principle of invariance helps in reducing the complexity of RTEs by allowing the separation of the medium into layers without changing the overall transfer characteristics. This principle implies that the propagation of radiation through a medium can be understood incrementally, layer by layer, which is particularly advantageous in numerical computations and analytical solutions. Mishchenko et al. (1999) considered an iterative solution of Eq. (4.58) in the discrete ordinate space, while Afanas'ev et al. (2013) found solutions by using the small-angle approximation.

Ambartsumyan's equations of invariance have found widespread applications in astrophysics, particularly in the study of the atmospheres of stars and planets. By applying these equations, researchers can more effectively model how light interacts with and escapes from these atmospheres.

4.3.2 *H*-functions

The problem of diffuse reflection can be reduced to finding the so-called H-functions. Let us apply Eq. (4.58) to the case of isotropic scattering so that the azimuth dependency vanishes. Equation (4.58) can be rewritten as follows:

$$R_\infty(\mu,\mu_0) = \frac{\omega_0}{4(\mu+\mu_0)}\left[1 + 2\mu_0\int_0^1 R_\infty(\mu',\mu)d\mu' + 2\mu\int_0^1 R_\infty(\mu',\mu_0)d\mu' \right. \\ \left. + 4\mu\mu_0\int_0^1 d\mu' R_\infty(\mu',\mu_0)\int_0^1 d\mu'' R_\infty(\mu'',\mu_0)\right]. \tag{4.60}$$

This equation can be rearranged as follows:

$$R_\infty(\mu,\mu_0) = \frac{\omega_0 H(\mu)H(\mu_0)}{4(\mu+\mu_0)}, \tag{4.61}$$

where

$$H(\mu) = 1 + 2\mu\int_0^1 R_\infty(\mu,\mu_0)d\mu_0. \tag{4.62}$$

The substitution of Eq. (4.62) to Eq. (4.61) gives the following nonlinear integral equation for H-functions:

$$H(\mu) = 1 + \frac{1}{2}\mu H(\mu)\omega_0\int_0^1 \frac{H(\mu')}{\mu+\mu'}d\mu'. \tag{4.63}$$

The advantage of using H-functions is that the calculation of the reflection function for isotropic scattering is reduced to the determination of the function of one variable.[1] H-functions have several interesting mathematical properties (Chandrasekhar, 1950; Nagirner and Ivanov, 2020).

Additionally, an analytical solution to this integral equation is available, as detailed by Fok (1944):

$$H(\eta) = \exp(-\alpha\eta), \tag{4.64}$$

where α is defined by the following equation:

$$\alpha = \frac{1}{\pi}\int_0^\infty \frac{\ln z(x)\,dx}{1+\eta^2x^2}, \tag{4.65}$$

---

1 Several problems of radiative transfer were reduced to H-functions. In particular, the possibility of reduction to H-functions for azimuth-averaged radiation fields and arbitrary phase functions was found by Kuščer (1955). The problem in the context of H-functions for semi-infinite media with sources at infinity or finite depth was solved by Kourganoff (1952).

and $z(x)$ is given by

$$z(x) = 1 - \omega \frac{\arctan x}{x}. \tag{4.66}$$

Several analytical approximations for this function exist. Notably, Hapke (1993) proposed the following approximation:

$$H(\eta) = \frac{1 + 2\eta}{1 + 2\eta\sqrt{1 - \omega}}. \tag{4.67}$$

This approximation yields an error of less than 5%.

4.3.3 Nonabsorbing Media

From the expression for the reflection function of a semi-infinite layer with isotropic scattering (see Eq. (4.61)), it follows that the determination of the reflection function of a semi-infinite medium is reduced to the calculation of integral given by Eq. (4.65) required to evaluate H-functions. Let us consider the case of nonabsorbing media. In this case, the following approximation for Eq. (4.64) can be used:

$$H_0(\xi) = 1 + 2\xi \tag{4.68}$$

and, therefore, it follows from Eq. (4.61) for the reflection function of a semi-infinite nonabsorbing turbid layer with isotropic scattering:

$$R(\xi, \eta) = \frac{(1 + 2\xi)(1 + 2\eta)}{4(\xi + \eta)}. \tag{4.69}$$

Equation (4.69) can be generalized for the case of anisotropic scattering. First of all, one can see that the reflection function can be presented as the sum of the single- and multiple light scattering terms:

$$R(\xi, \eta) = R_{ss}(\xi, \eta) + R_{ms}(\xi, \eta), \tag{4.70}$$

where

$$R_{ss}(\xi, \eta) = 0.25(\xi + \eta)^{-1}, \tag{4.71}$$

$$R_{ms}(\xi, \eta) = \frac{\xi + \eta + 2\xi\eta}{2(\xi + \eta)}. \tag{4.72}$$

We suggest that Eq. (4.70) can be used for anisotropically light scattering media as well. It follows in this case (see the single-scattering solution Eq. (3.19)):

$$R_{ss}(\xi, \eta) = 0.25p(\theta)(\xi + \eta)^{-1} \tag{4.73}$$

and we suggest to use the following approximation for $R_{ms}(\xi, \eta)$:

$$R_{ms}(\xi, \eta) = \frac{\Gamma_1(\xi + \eta) + \Gamma_2 \xi \eta + \Gamma_3}{2(\xi + \eta)} \tag{4.74}$$

and the constants $\Gamma_j (j = 1, 2, 3)$ can be found using the polynomial approximation for the function

$$D(\xi) \equiv 4\xi R_{ms}(\xi, \eta) = \Gamma_3 + 2\Gamma_1 \xi + \Gamma_2 \xi^2 \tag{4.75}$$

and the constants Γ_j $(j = 1, 2, 3)$ can be derived from fitting radiative transfer calculations to this equation. It follows from the aforementioned discussion that $\Gamma_1 = \Gamma_2 = 1, \Gamma_3 = 0$ for isotropic scattering. We have derived for the fractal ice grain phase function: $\Gamma_1 = 0.593$, $\Gamma_2 = 2.579$, and $\Gamma_3 = 0.624$. Therefore, we propose the following approximation for the reflectance for SSA equal to unity (conservative scattering case):

$$R_0 = \frac{\Gamma_1(\xi + \eta) + \Gamma_2 \xi \eta + \Gamma_3 + p(\theta)/2}{2(\xi + \eta)}, \tag{4.76}$$

where $p(\theta)$ is the phase function. The reflection function calculated using the numerical solution of the RTE for a semi-infinite turbid medium composed of nonabsorbing spherical water droplets (water cloud) and ice fractal particles (snow) is shown in Figure 4.3 as a function of the incident angle for nadir observation at 550 nm. The spherical water droplets are characterized by the effective droplet radius 6 microns and the coefficient of variance of gamma size distribution 38%. The phase function of the fractal particles has been calculated

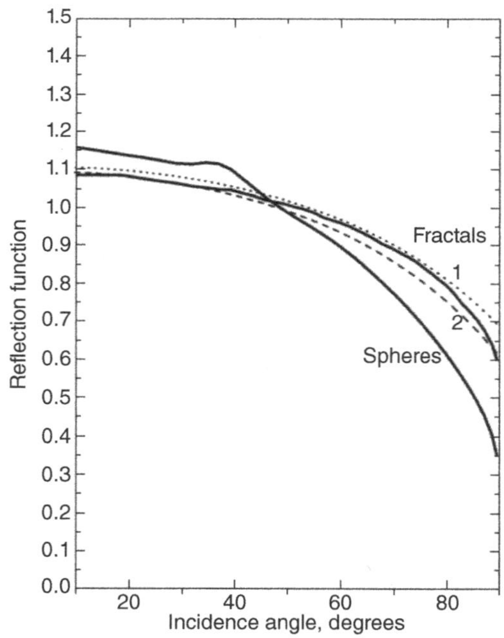

Figure 4.3 Intercomparison of reflection functions of nonabsorbing semi-infinite media composed of spherical and nonspherical scatterers. *Source:* Adapted from Kokhanovsky (2006).

using the fractal model of ice grains (Macke et al., 1996). The dashed line 1 in Figure 4.3 corresponds to Eq. (4.76). The dashed line 2 in Figure 4.3 corresponds to Eq. (4.76) with the assumption that the last term in the nominator can be neglected. It follows that Eq. (4.76) can be used to describe the angular behavior of the snow reflection function in channels where light absorption effects by snow can be neglected. One can see that the reflection function for the case of fractal snow grains is smooth and closer to the Lambertian case ($R = 1$) as compared to the water cloud model. Also, there is no peak for the reflectance of turbid media with fractal particles in the rainbow region characteristic of water droplets. Therefore, one can conclude that the assumption of spherical ice particles can lead to substantial errors in the interpretation of angular intensity dependence for the case of light reflected from snow layers. This is not so crucial for the light transmitted by thick snow layers because of randomization of photon direction.

4.3.4 Weakly Absorbing Media

Let us consider now the case of weakly absorbing media. In this case, the SSA is close to 1, and one expects that the reflection function of a semi-infinite turbid layer can be calculated using the correction of the results for the case of SSA equal to 1.0. Indeed, this is the case, and reflection function of a semi-infinite turbid layer can be presented as (Sobolev, 1975):

$$R_\infty(\omega_0) = R_{0\infty} - y(\omega_0)\, u_0(\mu) u_0(\mu_0), \tag{4.77}$$

where it is assumed that $\omega_0 \to 1$ and

$$y(\omega_0) = 4\sqrt{\frac{1-\omega_0}{3(1-g)}}. \tag{4.78}$$

Here, g is the asymmetry parameter, and the function $u_0(\mu)$ can be calculated using the following expression (Kokhanovsky, 2021):

$$u_0(\mu) = 0.6\mu + (1 + \sqrt{\mu})/3. \tag{4.79a}$$

Alternatively, one can use the following approximation (Sobolev, 1975):

$$u_0(\mu) = 3(1 + 2\mu)/7. \tag{4.79b}$$

Equation (4.79a) is more accurate at a small value of μ, as demonstrated in Figure 4.4. The escape function calculated using exact radiative transfer theory for the Henyey–Greenstein phase function at $g = 0.75$ is given in Table 4.1.

One can see that the calculation of the reflection function for weakly absorbing medium is reduced to the case of a nonabsorbing semi-infinite turbid medium studied in the previous section. Equation (4.77) can be applied only in the narrow range of the value of single-scattering albedo (SSA) close to one. To extend the applicability of the theory, Zege et al. (1991) proposed to use the exponential approximation. Specifically, Eq. (4.77) is substituted by the following formula:

$$R_\infty(\omega_0) = R_{0\infty} \exp(-y u_0(\mu) u_0(\mu_0)/R_{0\infty}). \tag{4.80}$$

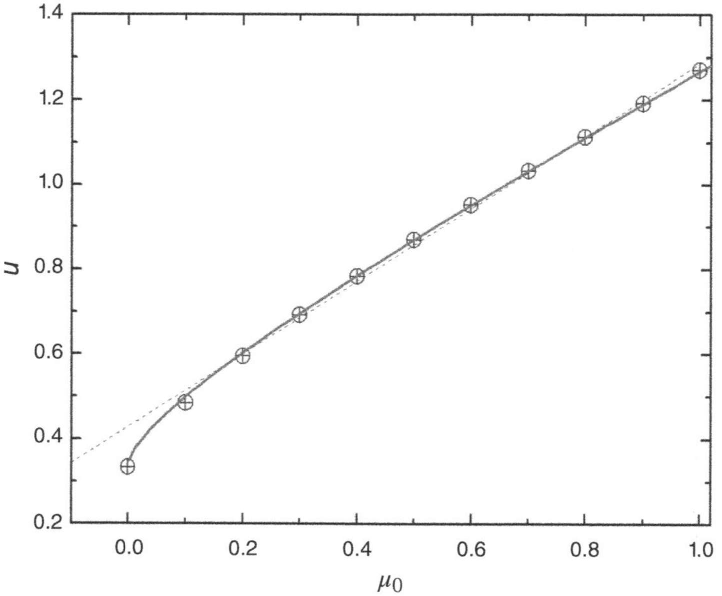

Figure 4.4 Dependence of the escape function on the cosine of the solar zenith angle for the Henyey–Greenstein phase function with $g = 0.75$, calculated using exact radiative transfer calculations (symbols), Eq. (4.79b) (dashed line), and Eq. (4.79a) (solid line).

Table 4.1 Escape function for the Henyey–Greenstein phase function with $g = 0.75$ (Yanovitskij, 1997).

μ_0	u_0
0.0	0.333
0.1	0.484
0.2	0.595
0.3	0.692
0.4	0.783
0.5	0.869
0.6	0.952
0.7	1.033
0.8	1.113
0.9	1.192
1.0	1.27

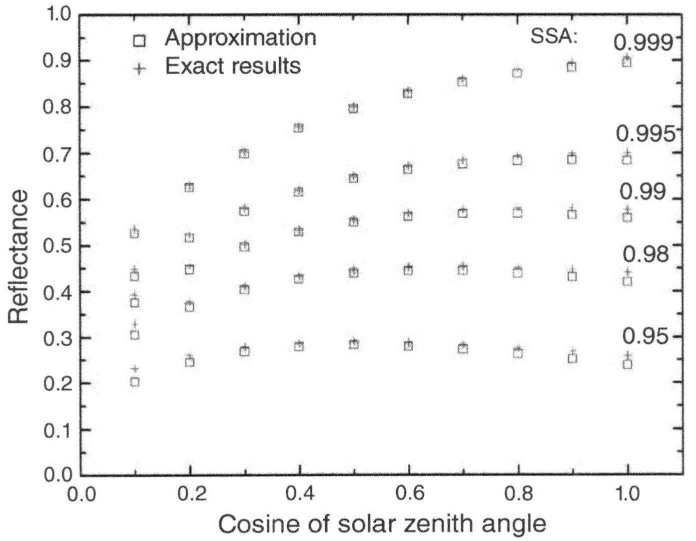

Figure 4.5 Dependence of the reflection function of a semi-infinite medium on the cosine of the incident zenith angle for various values of the single scattering albedo (SSA) (0.95–0.999) at nadir observation for the Henyey–Greenstein phase function with asymmetry parameter $g = 0.75$. The results derived from the numerical solution of the exact RTE (exact results) and from Eq. (4.80) (approximation) are shown. *Source:* Adapted from Kokhanovsky et al. (2019) (licensed under CC BY 4.0).

Equation (4.80) has also been derived by Kokhanovsky (2002) using the photon random walk model.

One can see that Eq. (4.79a) transforms to Eq. (4.77) as $\omega_0 \to 1$. However, it can be used for a larger range of SSA values as demonstrated in Figure 4.5.

5

Optically Thick Media

5.1 Nonabsorbing Media

Let us consider the case of finite layers now. Assuming that optical thickness is large, one can derive the following relationship between the reflection functions of a semi-infinite and finite layers having the same phase function and single-scattering albedo (Sobolev, 1975):

$$R_{0\infty} = R(\tau) + tu_0(\xi)u_0(\eta), \tag{5.1}$$

where the escape function has been considered in the previous section and t is the diffuse transmittance of the finite layer under diffuse illumination conditions. Equation (5.1) states that the reflection of a semi-infinite layer can be separated into two contributions:

- the reflection of light from upper optically thick, nonabsorbing layer;
- the contribution from semi-infinite layer below the optically thick layer, which is determined by the diffuse transmittance t and respective angular functions.

Taking into account the normalization condition (Sobolev, 1975):

$$2\int_0^1 u_0(\eta))\eta d\eta = 1, \tag{5.2}$$

and performing the integration of Eq. (5.1) with respect to the angular variables, we derived from Eq. (5.1):

$$r + t = 1, \tag{5.3}$$

which also follows from the energy conservation law for a finite layer. Here,

$$r = 4\int_0^1 \bar{R}(\tau,\xi,\eta)\xi\eta d\xi d\eta \tag{5.4}$$

is the spherical albedo and

$$\bar{R}(\tau,\xi,\eta) = (2\pi)^{-1}\int_0^{2\pi} R(\tau,\xi,\eta,\varphi) \tag{5.5}$$

is the azimuthally averaged phase function.

Analytical Methods in Radiative Transfer, First Edition. Alexander Kokhanovsky, Vijay Natraj and Dmitry Efremenko.

An important point is that the reflection function of an optically thick, nonabsorbing turbid layer can be presented via the reflection function of a nonabsorbing semi-infinite layer $R_{0\infty}$ studied earlier. Specifically, it follows from Eq. (5.1):

$$R(\tau) = R_{0\infty} - tu_0(\xi)u_0(\eta), \tag{5.6}$$

where the only unknown parameter is the diffuse transmittance coefficient t. This parameter depends on the value of the optical thickness τ and also on the asymmetry parameter g. Let us show it.

First, we note that the diffuse transmittance coefficient $t(\tau)$ of the combination of two optically thick layers with the optical thicknesses τ_1 and τ_2 can be presented as

$$t(\tau) = \frac{t(\tau_1)t(\tau_2)}{1 - r(\tau_1)r(\tau_2)}, \tag{5.7}$$

where $\tau = \tau_1 + \tau_2$. Taking into account Eq. (5.3), we derive

$$t(\tau) = \frac{t(\tau_1)t(\tau_2)}{t(\tau_1) + t(\tau_2) - t(\tau_1)t(\tau_2)}. \tag{5.8}$$

It follows from Eq. (5.8)

$$t^{-1}(\tau_1 + \tau_2) - 1 = t^{-1}(\tau_1) - 1 + t^{-1}(\tau_2) - 1. \tag{5.9}$$

It follows from Eq. (5.9) that $t^{-1}(\tau) - 1$ is the linear function of τ:

$$t^{-1}(\tau) - 1 = c\tau. \tag{5.10}$$

The parameter c does not depend on the optical thickness but has the following dependence on the asymmetry parameter g (Sobolev, 1975):

$$c = 0.75(1 - g). \tag{5.11}$$

Therefore, we derive

$$t(\tau) = \frac{1}{1 + 0.75(1 - g)\tau}. \tag{5.12}$$

Equations (5.6)–(5.12) can be used to derive the reflection function of a semi-infinite nonabsorbing layer. More accurate analysis with the usage of the radiative transfer equation (RTE) (Sobolev, 1975; Kokhanovsky, 2006) shows that the value of 1 must be substituted by 1.072 in Eq. (5.12) for typical geophysical phase functions.

Due to angular relaxation of photons and the reciprocity principle, the transmission function of an optically thick, nonabsorbing turbid layer can be presented as

$$T(\tau) = tu_0(\xi)u_0(\eta) \tag{5.13}$$

and, therefore,

$$R(\tau) = R_{0\infty} - T(\tau) \tag{5.14}$$

or

$$R_{0\infty} = R(\tau) + T(\tau) \tag{5.15}$$

for the case under study.

5.2 Weakly Absorbing Media

We shall assume that the separation of the angular and optical thickness variables as shown in Eq. (5.13) is valid not only for nonabsorbing but also for weakly absorbing media. Therefore, the following relationship holds:

$$T(\xi,\eta,\tau) = t(\tau)u_0(\xi)u_0(\eta). \tag{5.16}$$

However, Eq. (5.6) must be modified to take into account additional light absorption:

$$R_{0\infty} = R(\tau) + qtu_0(\xi)u_0(\eta). \tag{5.17}$$

The parameter q can be estimated if the pair (r, t) is known. Indeed, integrating Eq. (5.17) with respect to the angular variables, we have

$$q = \frac{r_\infty - r}{t}. \tag{5.18}$$

Let us derive analytical approximation for the parameters (r, t) assuming that light can propagate only along a line (1D case). Clearly, the reflected light fraction is given by the following expression in this simplified case:

$$r = \alpha d\tau, \tag{5.19}$$

where $\alpha = (1-f)\omega_0$, $1-f$ is the probability of light scattering in the backward direction, and ω_0 is the probability of photon survival in the light scattering event. Let us establish now the differential equation, which governs the dependence of r on the value of the optical thickness τ. For this, we use Eq. (2.88) derived by Stokes for two slabs:

$$r(\tau + d\tau) = r(\tau) + \frac{\alpha t^2(\tau)\, d\tau}{1 - r(\tau)\,\alpha d\tau}, \tag{5.20}$$

with the slab having the optical thicknesses τ and $d\tau$, respectively. This equation can be simplified by neglecting the terms of the second order and higher with respect to the value of $d\tau$. Specifically, it follows (Ambartsumian, 1944a):

$$r'(\tau) = \alpha t^2(\tau). \tag{5.21}$$

We also note that Eq. (2.88) gives for the reflectance of a semi-infinite homogeneous turbid layer:

$$r_\infty = r(\tau) + \frac{t^2(\tau)\, r_\infty}{1 - r(\tau)\, r_\infty}, \tag{5.22}$$

where τ is the optical thickness of a finite turbid layer attached to the semi-infinite layer from above. It is assumed that optical properties of the upper and lower layers are the same. This means that

$$t^2(\tau) = (r - r_\infty)(r - r_\infty^{-1}). \tag{5.23}$$

Therefore, we arrive to the following differential equation for the reflectance:

$$r'(z) = (r - r_\infty)(r - r_\infty^{-1}), \tag{5.24}$$

where $z = \alpha\tau$. This equation has the following analytical solution (Kamke, 1959):

$$r(z) = r_\infty \frac{1 - \exp(-\nu z)}{1 - r_\infty^2 \exp(-\nu z)}, \tag{5.25}$$

where $\nu = r_\infty^{-1} - r_\infty$. This equation can be rewritten as

$$r(\tau) = r_\infty \frac{1 - \exp(-2k\tau)}{1 - r_\infty^2 \exp(-2k\tau)}, \tag{5.26}$$

where

$$k = \frac{1}{2}(1 - f)(r_\infty^{-1} - r_\infty)\omega_0. \tag{5.27}$$

The value of transmittance can be derived from Eqs. (5.23)–(5.26). The result is

$$t(\tau) = \frac{(1 - r_\infty^2)\exp(-k\tau)}{1 - r_\infty^2 \exp(-2k\tau)}. \tag{5.28}$$

It follows from Eq. (4.80) (Zege et al., 1991) $r_\infty = \exp(-y)$ for weakly absorbing media and, therefore, the following relationships hold:

$$r(\tau) = \frac{\exp(-y) - \exp(-2x - y)}{1 - \exp(-2x - 2y)}, \tag{5.29a}$$

$$t(\tau) = \frac{(1 - \exp(-2y))\exp(-x)}{1 - \exp(-2x - 2y)} \tag{5.29b}$$

or, alternatively,

$$r(\tau) = \frac{\mathrm{sh} x}{\mathrm{sh}(x + y)}, \tag{5.30}$$

$$t(\tau) = \frac{\mathrm{sh} y}{\mathrm{sh}(x + y)}, \tag{5.31}$$

where

$$x = k\tau, \tag{5.32}$$

$$y = 4\sqrt{\frac{1 - \omega_0}{3(1 - g)}}. \tag{5.33}$$

The value of k can be presented in the following way as $\omega_0 \to 1$ and $r_\infty \to 1 - y$:

$$k = (1 - f)y. \tag{5.34}$$

It follows from Eqs. (5.31) and (5.34) at $y = 0$ (no absorption):

$$t(\tau) = \frac{1}{1 + (1 - f)\tau}, \tag{5.35}$$

Comparing Eqs. (5.35) and (5.12), we derive

$$1 - f = 3(1 - g)/4. \tag{5.36}$$

Then it follows from Eq. (5.34) that

$$k = \sqrt{3(1-\omega_0)(1-g)}. \tag{5.37}$$

Finally, we derive the following analytical approximations for the reflection and transmission functions (Zege et al., 1991):

$$R(\tau) = R_\infty - \frac{\mathrm{sh}(y)}{\mathrm{sh}(x+y)} \exp(-x-y) u(\mu_0) u(\mu), \tag{5.38}$$

$$T(\mu_0, \mu) = \frac{\mathrm{sh}(y)}{\mathrm{sh}(x+y)} u_0(\mu_0) u_0(\mu), \tag{5.39}$$

where we have accounted for the fact that $q = \exp(-x-y)$ (see Eqs. (5.18), (5.30), and (5.31)).

5.3 Optically Thick Turbid Media with an Arbitrary Level of Light Absorption

We now extend our analysis to optically thick turbid media with arbitrary absorption levels. Before deriving analytical expressions for the reflection and transmission functions, R and T, in the asymptotic limit $\tau_0 \gg 1$, it is necessary to establish some auxiliary relations that hold in deep layers of such media.

A well-known property of optically thick media is that, at sufficient depth, the intensity of light becomes independent of the azimuthal angle. Under this condition, the RTE, integrated over azimuthal angles, takes the following form (see Section 2.9 for details):

$$\eta \frac{dI(\tau,\eta)}{d\tau} = -I(\tau,\eta) + B(\tau,\eta) + B_0(\tau,\eta), \tag{5.40}$$

where the terms $B(\tau,\eta)$ and $B_0(\tau,\eta)$ account for scattered radiation and the contribution from the external incident flux, respectively:

$$B(\tau,\eta) = \frac{\omega_0}{2} \int_{-1}^{1} p(\eta,\eta') I(\tau,\eta') \, d\eta', \tag{5.41}$$

$$B_0(\tau,\eta) = \frac{\omega_0 F}{4\pi} p(\eta,\xi) e^{-\tau/\xi}, \tag{5.42}$$

where $p(\eta, \xi)$ represents the azimuthally averaged phase function:

$$p(\eta,\xi) = \frac{1}{2\pi} \int_0^{2\pi} p(\eta,\xi,\varphi) \, d\varphi. \tag{5.43}$$

In this formulation, polarization effects are neglected for simplicity. These equations provide the foundation for studying radiative transfer in strongly scattering and absorbing media, where multiple scattering dominates and the angular dependence of intensity evolves toward asymptotic behavior.

Let us assume that $\tau \to \infty$. Then it follows that $B_0(\tau, \eta) \to 0$ and

$$I(\tau, \eta) = Ci(\eta)\, e^{-k\tau}, \tag{5.44}$$

where C is a constant. It is important to note that Eq. (5.40) remains valid for any constant C in the asymptotic limit $\tau \to \infty$. This equation describes the so-called deep-layer regime, where the parameters η and τ become effectively decoupled. In this regime, the total light flux f decreases progressively with increasing optical depth while maintaining the angular distribution of scattered radiation, which is characterized by the function $i(\eta)$. The optical depth at which the flux f is attenuated by a factor of e is given by $\tau_e = 1/k$, where k is the diffusion exponent. Both the angular distribution function $i(\eta)$ and the exponent k play a crucial role in the theoretical framework considered here. These parameters not only define the behavior of radiation in the deep-layer regime but also determine the intensity of transmitted and reflected light, as will be demonstrated in subsequent sections.

We have from Eq. (5.40):

$$(1 - k\eta)\, i(\eta) = \frac{\omega_0}{2} \int_{-1}^{1} p(\eta, \eta')\, i(\eta')\, d\eta'. \tag{5.45}$$

This integral equation is usually solved numerically. Let us assume that $p = 1$, i.e. isotropic scattering. Then we have the following analytical solution:

$$i(\eta) = \frac{\omega_0}{2(1 - k\eta)} \int_{-1}^{1} i(\eta')\, d\eta' \tag{5.46}$$

or

$$i(\eta) = \frac{D}{1 - k\eta}, \tag{5.47}$$

where

$$D = \frac{\omega_0}{2} \int_{-1}^{1} i(\eta')\, d\eta' \tag{5.48}$$

does not depend on the angle. Note that $i(\eta)$ satisfies the RTE (5.45) for any constant D and, therefore,

$$i(\eta) = \frac{1}{1 - k\eta} \tag{5.49}$$

and

$$I(\tau, \eta) = \frac{\exp(-k\tau)}{1 - k\eta}, \tag{5.50}$$

where we used the normalization condition $D = 1$. The diffusion constant k can be determined as follows. Substituting Eq. (5.47) into Eq. (5.46) and considering that $D = 1$, we obtain

$$\frac{\omega_0}{2k} \ln\left(\frac{1 + k}{1 - k}\right) = 1. \tag{5.51}$$

This equation provides a means to determine the diffusion constant k for arbitrary values of the single-scattering albedo ω_0 under the assumption that the phase function is isotropic, i.e. $p(\theta) = 1$. In the special case, where $\omega_0 = 1$, it follows that $k = 0$, indicating that no absorption occurs, and the radiation field remains fully diffusive. The results of the corresponding calculations for the Henyey–Greenstein phase function,

$$p(\theta) = \sum_{n=1}^{\infty} g^n P_n(\cos\theta) \tag{5.52}$$

for asymmetry parameters $g = 0.75$ (ice clouds) and $g = 0.85$ (water clouds) are given in Figure 5.1 (Yanovitskij, 1997). From this result, it follows that $0 \leq k \leq 1$. This implies that in an absorbing medium, the attenuation of diffuse light in the deep-layer regime, given by $\exp(-k\tau)$, occurs at a slower rate compared to the attenuation of direct (unscattered) light, which follows $\exp(-\tau)$. Consequently, multiple scattering effectively prolongs the persistence of radiation within the medium, reducing the rate at which light is absorbed.

It also follows that (see Eq. (5.45)):

$$(1 + k\eta)\, i(-\eta) = \frac{\omega_0}{2} \int_{-1}^{1} p(-\eta, \eta')\, i(\eta')\, d\eta' \tag{5.53}$$

or

$$(1 + k\eta)\, i(-\eta) = \frac{\omega_0}{2} \int_{-1}^{1} p(\eta, \eta')\, i(-\eta')\, d\eta', \tag{5.54}$$

where we used the property: $p(-\eta, -\eta') = p(\eta, \eta')$.

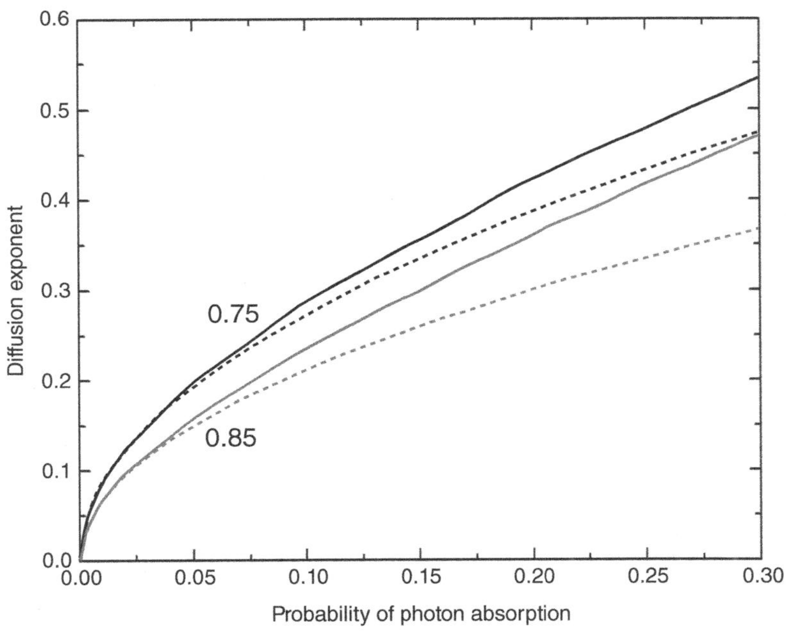

Figure 5.1 Dependence of diffusion exponent on the probability of photon absorption $\beta = 1 - \omega_0$ for the Henyey–Greenstein phase function with $g = 0.75$ and 0.85 (solid lines) (Yanovitskij, 1997). Results derived using the approximation $k = \sqrt{3\beta(1-g)}$ that is valid for small values of the probability of light absorption are also shown (dashed lines).

Next, we establish the relationship between the downward-propagating intensity $i^{\downarrow}(\eta)$ and the upward-propagating intensity $i^{\uparrow}(-\eta)$. The arrows indicate the direction of light propagation. By definition, the quantity $\eta = |\cos\vartheta|$ is always positive, while negative values of η correspond to upwelling light flux. To derive this relationship, we consider a horizontal plane located at a large optical depth $\tau \gg 1$, far from the boundaries. This plane is illuminated not only by radiation coming from above, with intensity i_a, but also by radiation originating from below. The latter component, denoted as i_b, represents light that has been reflected from the overlying layers and contributes to the total intensity at the given depth. Clearly, we have:

$$i^{\downarrow}(\eta) = i_a(\eta) + i_b(\eta). \tag{5.55}$$

Thus, the function $i^{\downarrow}(\eta)$ can be expressed as the sum of two contributions. The first term, $i_a(\eta)$, corresponds to the radiation directly transmitted from the upper layer, while the second term accounts for the radiation reflected from the overlying medium. Since $i_a(\eta)$ represents light that has passed through the upper layer, it is naturally proportional to the angular distribution function $u(\eta)$, which characterizes the transmission properties of the medium:

$$i_a(\eta) = Mu(\eta), \tag{5.56}$$

where M is the unknown proportionality constant. We will find this constant at later stages of our derivations. Also, it follows for the intensity $i_b(\eta)$ that

$$i_b(\eta) = 2\int_0^1 R(\eta,\eta')\, i(-\eta')\,\eta' d\eta', \tag{5.57}$$

where $R(\eta,\eta')$ is the azimuthally averaged reflection function of the upper layer under illumination from below ($\eta > 0, \eta' > 0$). The thickness of this layer can be chosen arbitrarily. For simplicity, we assume that the reflection function $R(\eta,\eta')$ is equivalent to the azimuthally averaged reflection function of a semi-infinite medium, denoted as $R_\infty(\eta,\eta')$.

Summing up, it follows that:

$$i^{\downarrow}(\eta) = Mu(\eta) + 2\int_0^1 R_\infty(\eta,\xi)\, i(-\xi)\,\xi d\xi. \tag{5.58}$$

To determine M, we multiply the last equation by $\eta i^{\downarrow}(\eta)$ and integrate over the range $0 \le \eta \le 1$. This yields

$$\int_0^1 \eta i^{\downarrow 2}(\eta)\, d\eta = M\int_0^1 u(\eta) i^{\downarrow}(\eta)\,\eta d\eta + \Im, \tag{5.59}$$

where the two-dimensional integral

$$\Im = 2\int_0^1 i^{\downarrow}(\eta)\,\eta d\eta \int_0^1 R_\infty(\eta,\xi)\, i^{\uparrow}(-\xi)\,\xi d\xi \tag{5.60}$$

can be simplified. For this, we note that (see Eq. (5.57))

$$i^{\uparrow}(-\xi) = 2\int_0^1 i^{\downarrow}(\eta) R_\infty(\eta,\xi)\,\eta d\eta \tag{5.61}$$

and, therefore,

$$\mathfrak{I} = \int_0^1 i^{\uparrow 2}(-\xi)\,\xi d\xi \tag{5.62}$$

or

$$\mathfrak{I} = -\int_{-1}^0 i^{\downarrow 2}(\xi)\,\xi d\xi. \tag{5.63}$$

Therefore, it follows that

$$M = \Lambda \int_{-1}^1 i^2(\eta)\,\eta d\eta, \tag{5.64}$$

where

$$\Lambda = \left[\int_0^1 u(\eta)\, i(\eta)\, d\eta\right]^{-1}. \tag{5.65}$$

We will use the normalization condition $\Lambda = 2$. Then one can derive

$$M = 2\int_{-1}^1 i^2(\eta)\,\eta d\eta. \tag{5.66}$$

We present the equation for M together with other important relationships in Table 5.1.

Table 5.1 Main equations and constants.

	Property
1	$(1-k\eta)\, i(\eta) = \frac{\omega_0}{2}\int_{-1}^1 p(\eta,\eta')\, i(\eta')\, d\eta'$
2	$(1+k\eta)\, i(-\eta) = \frac{\omega_0}{2}\int_{-1}^1 p(\eta,\eta')\, i(-\eta')\, d\eta'$
3	$\frac{\omega_0}{2}\int_{-1}^1 i(\eta)\, d\eta = 1$
4	$i(-\eta) = 2\int_0^1 i(\xi)\, R_\infty(\xi,\eta)\,\xi d\xi$
5	$i(\eta) = 2\int_0^1 i(-\xi)\, R_\infty(\xi,\eta)\,\xi d\xi + Mu(\eta)$
6	$2\int_0^1 u(\eta)\, i(\eta)\,\eta d\eta = 1$
7	$M = 2\int_{-1}^1 i^2(\eta)\,\eta d\eta$
8	$N = 2\int_0^1 i(-\eta)\, K(\eta)\,\eta d\eta$
9	$\delta = \int_0^1 u_0(\xi)\,\xi^2 d\xi$
10	$u_0(\xi) = \frac{3}{4}\left(\xi + 2\int_0^1 R_{0\infty}(\xi,\eta)\,\eta^2 d\eta\right)$
11	$2\int_0^1 u_0(\eta)\,\eta d\eta = 1$
12	$2\int_0^1 R_{0\infty}(\xi,\eta)\,\eta d\eta = 1$

The constant N defined in Property 8 in Table 5.1 will be used in further derivations devoted to the studies of relationships between auxiliary functions defined as follows:

$$P(\tau) = \int_{-1}^{1} \eta i(\eta) I(\tau, \eta)\, d\eta \tag{5.67}$$

and

$$Q(\tau) = \int_{-1}^{1} \eta i(-\eta) I(\tau, \eta)\, d\eta. \tag{5.68}$$

The relationship between the functions $P(\tau)$ and $Q(\tau)$ provides a foundation for deriving asymptotic equations for the reflection and transmission functions in the limit $\tau_0 \to \infty$. In the following, we apply the approach developed by Sobolev (1975) to establish these asymptotic expressions.

First, we note that it follows from Eq. (5.45) after multiplication of this equation by $i(\eta)$ and integration from -1 to 1:

$$\frac{dP(\tau)}{d\tau} = -kP(\tau) + P_0(\tau), \tag{5.69}$$

where

$$P_0(\tau) = \int_{-1}^{1} i(\eta) B_0(\tau, \eta)\, d\eta \tag{5.70}$$

and we use the equality:

$$-kP(\tau) = \int_{-1}^{1} B(\tau, \eta) i(\eta)\, d\eta - \int_{-1}^{1} i(\eta) I(\tau, \eta)\, d\eta. \tag{5.71}$$

This equality follows directly from Property 2 in Table 5.1. By multiplying Property 1 in Table 5.1 by $I(\tau, \eta)$ and integrating over the range $-1 \le \eta \le 1$, we obtain

$$\int_{-1}^{1} I(\tau, \eta) i(\eta)\, d\eta - kP(\tau) = \frac{\omega_0}{2} \int_{-1}^{1} d\eta \int_{-1}^{1} I(\tau, \eta) p(\eta, \eta') i(\eta')\, d\eta' \tag{5.72}$$

or

$$\int_{-1}^{1} I(\tau, \eta) i(\eta)\, d\eta - kP(\tau) = \int_{-1}^{1} d\eta' B(\tau, \eta') i(\eta')\, d\eta', \tag{5.73}$$

where we use the property $p(\eta, \eta') = p(\eta', \eta)$. The last equation coincides with Eq. (5.71).

The next step is to find $P(\tau)$. For this, we use the following substitution in Eq. (5.69):

$$P(\tau) = f(\tau) e^{-k\tau}. \tag{5.74}$$

Then it follows that

$$\frac{df(\tau)}{d\tau} = P_0(\tau) e^{k\tau}. \tag{5.75}$$

This means that

$$f(\tau) = f(\tau_1) + \int_{\tau_1}^{\tau} P_0(t) e^{kt} dt. \tag{5.76}$$

Therefore, we have

$$P(\tau) = f(\tau_1)e^{-k\tau} + e^{-k\tau}\int_{\tau_1}^{\tau} P_0(t)e^{kt}dt. \tag{5.77}$$

The value of τ_1 can be found from boundary conditions. In particular, we are interested in the diffuse light. Diffuse light does not enter the medium from above or below ($I(0,\eta) = 0$ for $\eta > 0$ and $I(\tau_0,\eta) = 0$ for $\eta < 0$). Therefore, we have $\tau_1 = 0$. Then the boundary condition at the upper boundary is satisfied. Finally, it follows that

$$P(\tau) = P(0)e^{-k\tau} + \int_0^{\tau} P_0(t)e^{k(t-\tau)}dt. \tag{5.78}$$

A similar relationship can be obtained for $Q(\tau,\eta)$. Then we have

$$\frac{dQ(\tau)}{d\tau} = kQ(\tau) + Q_0(\tau), \tag{5.79}$$

where

$$Q_0(\tau) = \int_{-1}^{1} i(-\eta)B_0(\tau,\eta)\,d\eta. \tag{5.80}$$

This equation differs for the corresponding equation for P only by the sign before k. Therefore, we have

$$Q(\tau,\eta) = \psi(\tau_1^*)e^{k\tau} + e^{k\tau}\int_{\tau_1^*}^{\tau} Q_0(t)e^{-kt}dt, \tag{5.81}$$

where it was assumed that

$$Q(\tau) = \psi(\tau)e^{k\tau}. \tag{5.82}$$

The value of τ_1^* can be found from the boundary condition at the lower boundary of a medium. Namely, we have $\tau_1^* = \tau_0$. Therefore, it follows that

$$Q(\tau) = Q(\tau_0)e^{k(\tau-\tau_0)} + \int_{\tau_0}^{\tau} Q_0(t)e^{-k(t-\tau)}dt. \tag{5.83}$$

This equation is satisfied automatically at $\tau = \tau_0$.

Summing up, we have the following important relationships:

$$P(\tau) = P(0)e^{-k\tau} + V(\tau), \tag{5.84}$$

$$Q(\tau) = Q(\tau_0)e^{k(\tau-\tau_0)} + W(\tau), \tag{5.85}$$

where

$$V(\tau) = \int_0^{\tau} P_0(t)e^{k(t-\tau)}dt, \tag{5.86}$$

$$W(\tau) = \int_{\tau_0}^{\tau} Q_0(t)e^{-k(t-\tau)}dt. \tag{5.87}$$

These fundamental relationships hold for any optical depth τ and for any type of light source, as demonstrated by Sobolev (1975). They serve as a basis for deriving several key results in the optics of turbid media.

To proceed, we consider the specific case where $\tau = 0$ in the first equation and $\tau = \tau_0$ in the second equation. Under these conditions, we obtain

$$P(\tau_0) = P(0)\exp(-k\tau_0) + V(\tau_0), \tag{5.88}$$

$$Q(0) = Q(\tau_0)\exp(-k\tau_0) + W(0), \tag{5.89}$$

where

$$\begin{aligned} V(\tau_0, \eta) &= e^{-k\tau_0}\int_0^{\tau} P_0 dt \int_{-1}^{1} i(\eta)\frac{\omega_0}{4}p(\eta,\xi)\, e^{-t\left(\frac{1}{\xi}-k\right)} d\xi \\ &= \frac{1}{2}\left(e^{k\tau_0} - e^{-\frac{\tau_0}{\xi}}\right)\xi i(-\xi), \end{aligned} \tag{5.90}$$

and we use Property 2 in Table 5.1. Also, we have

$$\begin{aligned} W(0, \eta) &= \int_{\tau_0}^{0} e^{-kt} dt \int_{-1}^{1} i(-\eta)\frac{\omega_0}{4}p(\eta,\xi)\, e^{-t/\xi} d\eta \\ &= -\frac{1}{2}\left(1 - e^{-\left(k+\frac{1}{\xi}\right)\tau_0}\right)\xi i(-\xi). \end{aligned} \tag{5.91}$$

Therefore, neglecting small numbers $e^{-\tau_0/\xi}$, it follows as $\tau_0 \to \infty$:

$$P(0, \eta) = -P(\tau_0, \eta)\, e^{k\tau_0} + \frac{1}{2}\xi i(-\xi), \tag{5.92}$$

$$Q(0, \eta) = Q(\tau_0, \eta)\, e^{-k\tau_0} - \frac{1}{2}\xi i(-\xi). \tag{5.93}$$

These are auxiliary relations that can also be written in the following form:

$$i(\xi) = 2\int_{-1}^{1} T(\eta, \xi, \tau_0)\, i(\eta)\, \eta d\eta + 2e^{-k\tau_0}\int_{-1}^{1} R(\eta, \xi, \tau_0)\, i(-\eta)\, d\eta, \tag{5.94}$$

$$i(-\xi) = 2\int_{-1}^{1} R(\eta, \xi, \tau_0)\, i(\eta)\, \eta d\eta + 2e^{-k\tau_0}\int_{-1}^{1} T(\eta, \xi, \tau_0)\, i(-\eta)\, \eta d\eta, \tag{5.95}$$

where T and R are determined as follows:

$$R = \frac{\pi I_R}{\mu_0 F} \tag{5.96}$$

and

$$T = \frac{\pi I_T}{\mu_0 F}. \tag{5.97}$$

5.4 Asymptotic Equations

The general form of functions $R(\eta, \xi, \tau_0)$ and $T(\eta, \xi, \tau_0)$ can be obtained using physical arguments. In particular, T should be proportional to $u(\eta)$ (and, actually due to the reciprocity principle, also to $u(\xi)$). Therefore, we have (as discussed in Section 5.2):

$$T(\eta, \xi, \tau_0) = \alpha(\tau_0) u(\eta) u(\xi). \tag{5.98}$$

Let us now consider a semi-infinite layer and make a cut at a large optical thickness τ_0. Then we can represent $R_\infty(\eta, \xi)$ as a sum of reflection from the upper layer $R(\tau_0, \eta, \xi)$ and light transmitted by the upper layer and reflected back. The angular distribution of the transmitted light should be proportional to $u(\eta)$ as specified earlier. Therefore, we have

$$R_\infty(\eta, \xi) = R(\eta, \xi, \tau_0) + \beta(\tau_0) u(\eta) u(\xi). \tag{5.99}$$

Let us find α and β substituting the last two equations in the asymptotic Eqs. (5.94)–(5.95). Then it follows that

$$\begin{aligned} i(\xi) &= 2e^{k\tau_0} \int_{-1}^{1} \alpha(\tau_0) u(\eta) u(\xi) i(\eta) \eta d\eta \\ &+ 2 \int_{-1}^{1} (R_\infty(\eta, \xi) - \beta(\tau_0) u(\eta) u(\xi)) i(-\eta) \eta d\eta, \end{aligned} \tag{5.100}$$

$$\begin{aligned} i(-\xi) &= 2 \int_{-1}^{1} (R_\infty(\eta, \xi) - \beta(\tau_0) u(\eta) u(\xi)) i(\eta)\, d\eta \\ &+ 2e^{-k\tau_0} \int_{-1}^{1} \alpha(\tau_0) u(\eta) u(\xi) i(-\eta) \eta d\eta, \end{aligned} \tag{5.101}$$

and, therefore,

$$i(\xi) = e^{k\tau_0} \alpha(\tau_0) u(\xi) + i(\xi) - Mu(\xi) - \beta(\tau_0) u(\xi) N, \tag{5.102}$$

$$i(-\xi) = i(-\xi) - \beta(\tau_0) u(\xi) + \alpha N e^{-k\tau_0} u(\xi), \tag{5.103}$$

where we introduce the following integral:

$$N = 2 \int_{-1}^{1} u(\eta) i(-\eta) \eta d\eta. \tag{5.104}$$

It follows that

$$\alpha(\tau_0) - Me^{-k\tau_0} - \beta N e^{-k\tau_0} = 0, \tag{5.105}$$

$$\beta(\tau_0) = \alpha(\tau_0) N e^{-k\tau_0} \tag{5.106}$$

and, therefore,

$$\alpha(\tau_0) = \frac{Me^{-k\tau_0}}{1 - N^2 e^{-2k\tau_0}}. \tag{5.107}$$

Finally, we have

$$R(\eta, \xi, \varphi, \tau_0) = R_\infty(\eta, \xi, \varphi) - T(\xi, \eta, \tau_0) N e^{-k\tau_0}, \tag{5.108}$$

$$T(\xi, \eta) = \frac{Me^{-k\tau_0}}{1 - N^2 e^{-2k\tau_0}} u(\eta) u(\xi), \tag{5.109}$$

where we take into account the fact that the transmitted light does not depend on the azimuth φ. Equations (5.108) and (5.109) can be used for arbitrary phase functions and at any level of light absorption in the case of optically thick turbid media.

6

Turbid Media with Arbitrary Optical Thickness

6.1 Sobolev Approximation

The best results for layers having arbitrary optical thickness are obtained through the numerical solution of the radiative transfer equation (RTE). However, in many cases, it is desirable to have analytical approximations that are valid at any value of optical thickness. One such approximation has been developed by Sobolev (1975). Let us consider the Sobolev approximation for the case of a nonabsorbing turbid layer. In this case, the intensity of reflected and transmitted light depends on the optical thickness and the phase function. It has been demonstrated in the previous chapters that, in many cases, the phase function is of secondary importance and radiative transfer in turbid layers is governed by the value of the asymmetry parameter g. Therefore, we concentrate on the case of the linear (with respect to the cosine of the scattering angle) phase function:

$$p(\theta) = 1 + 3g\cos\theta. \tag{6.1}$$

The final expression, as we will see, will contain the parameters (g, τ) and the single-scattering phase function $p(\theta)$. The extension of this approximation to other phase functions can be done by substituting the linear phase function in the final expressions with the phase function valid for a given scattering medium. In addition, the approximations derived have better accuracy at small values of g. Therefore, their accuracy can be improved using a truncation procedure, as will be demonstrated later.

We start from the following RTE:

$$\cos\vartheta\frac{dI(\vartheta,\varphi)}{d\tau} + I(\vartheta,\varphi) - B(\vartheta,\varphi) = 0, \tag{6.2}$$

where

$$\begin{aligned} B(\vartheta,\varphi) = {} & \frac{\omega_0}{4\pi}\int_0^{2\pi} d\varphi' \int_{-1}^{1} d(\cos\vartheta')p(\Theta)\,I(\vartheta',\varphi') \\ & + \frac{\omega_0}{4}p(\Theta_0)\,F_0\exp\left(-\frac{\tau}{\cos\vartheta_0}\right), \end{aligned} \tag{6.3}$$

where

$$\cos\Theta = \cos\vartheta\cos\vartheta' - \sqrt{(1-\cos^2\vartheta)(1-\cos^2\vartheta')}\cos(\varphi-\varphi'), \tag{6.4}$$

Analytical Methods in Radiative Transfer, First Edition. Alexander Kokhanovsky, Vijay Natraj and Dmitry Efremenko.

$$\cos\Theta_0 = \cos\vartheta\cos\vartheta_0 - \sqrt{(1-\cos^2\vartheta)(1-\cos^2\vartheta_0)}\cos(\varphi-\varphi_0), \tag{6.5}$$

θ_0 and φ_0 are the incident zenith and azimuth angles, respectively. The boundary conditions are:

$$\begin{aligned} &I(\tau=0)=0,\ \cos\vartheta>0,\\ &I(\tau=\tau_0)=0,\ \cos\vartheta<0. \end{aligned} \tag{6.6}$$

They state that there is no diffuse light entering the plane–parallel turbid layer via its boundaries. The substitution of Eq. (6.1) to Eq. (6.3) gives:

$$B(\vartheta,\varphi) = B_1(\vartheta) + B_2(\vartheta)\cos(\varphi-\varphi_0), \tag{6.7}$$

where

$$B_1(\vartheta) = \sum_{s=1}^{3} b_s f_s(\tau), \tag{6.8}$$

$$B_2(\vartheta) = \sum_{s=1}^{2} c_s h_s(\tau), \tag{6.9}$$

$$\begin{aligned} b_1 &= 1,\\ b_2 &= 3g\eta,\\ b_3 &= F_0\,[1+b_2\xi]\,/4,\\ c_1 &= 3g\sqrt{1-\eta^2},\\ c_2 &= c_1F_0\sqrt{1-\xi^2}/4, \end{aligned} \tag{6.10}$$

$$\begin{aligned} f_1 &= \frac{1}{4\pi}\int_0^{2\pi} d\varphi' \int_{-1}^{1} I(\tau,\eta',\varphi')d\eta',\\ f_2 &= \frac{1}{4\pi}\int_0^{2\pi} d\varphi' \int_{-1}^{1} I(\tau,\eta',\varphi')\eta' d\eta',\\ f_3 &= h_2 = \exp(-\tau/\xi),\\ h_1 &= \frac{1}{4\pi}\int_0^{2\pi} \cos(\varphi'-\varphi_0)d\varphi' \int_{-1}^{1} I(\tau,\eta',\varphi')\sqrt{1-\eta'^2}d\eta'. \end{aligned} \tag{6.11}$$

Taking into account Eq. (6.7), we derive using Eq. (6.2):

$$I(\vartheta,\varphi) = I_1(\vartheta) + I_2(\vartheta)\cos(\varphi-\varphi_0), \tag{6.12}$$

where

$$\eta I_1'(\vartheta,\varphi) + I_1(\vartheta) - \bar{f}_1(\tau) - b_2\bar{f}_2(\tau) - b_3 f_3(\tau) = 0, \tag{6.13}$$

$$\eta I_2'(\vartheta,\varphi) + I_2(\vartheta) - c_1\bar{h}_1(\tau) - c_2 h_2(\tau) = 0 \tag{6.14}$$

and

$$\begin{aligned}
\bar{f}_1 &= \frac{1}{2}\int_{-1}^{1} I(\tau,\eta',\varphi')d\eta', \\
\bar{f}_2 &= \frac{1}{2}\int_{-1}^{1} I(\tau,\eta',\varphi')\eta' d\eta', \\
\bar{h}_1 &= \frac{1}{4}\int_{-1}^{1} I(\tau,\eta',\varphi')\sqrt{1-\eta'^2}d\eta'.
\end{aligned} \tag{6.15}$$

Let us derive approximate analytical equations for the functions given by Eq. (6.15). This makes it possible to solve Eqs. (6.13)–(6.14) analytically. Integrating Eq. (6.13) with respect to η, we derive:

$$\bar{f}_2'(\tau) - \frac{F_0}{4}\exp(-\tau/\xi) = 0 \tag{6.16}$$

or

$$\bar{f}_2(\tau) = -\frac{F_0}{4}\xi\exp(-\tau/\xi) + C_1, \tag{6.17}$$

where C_1 is the integration constant to be found from the boundary conditions.

Multiplying Eq. (6.13) by η and integrating with respect to η, we derive:

$$\Psi'(\tau) + (1-g)\bar{f}_2(\tau) - gF_0\xi\exp(-\tau/\xi)/4 = 0, \tag{6.18}$$

where

$$\Psi(\tau) = \frac{1}{2}\int_{-1}^{1} I_1(\eta')\eta'^2 d\eta' \tag{6.19}$$

and we assume that

$$\Psi(\tau) = \bar{f}_1(\tau)/3. \tag{6.20}$$

Then it follows from Eq. (6.18):

$$\bar{f}_1'(\tau) + 3(1-g)\bar{f}_2(\tau) - 3gF_0\xi\exp(-\tau/\xi)/4 = 0 \tag{6.21}$$

or

$$\bar{f}_1(\tau) = -3(1-g)C_1\tau - 3F_0\xi^2\exp(-\tau/\xi)/4 + C_2, \tag{6.22}$$

where the constants C_1 and C_2 can be found from the boundary conditions, which can be given in the following approximate form:

$$\begin{aligned}
\bar{f}_1(0) &= -2\bar{f}_2(0), \\
\bar{f}_1(\tau_0) &= 2\bar{f}_2(\tau_0).
\end{aligned} \tag{6.23}$$

Therefore, one can derive:

$$C_1 = \frac{F_0 \xi w(\xi, \tau_0)}{8(1 + 0.75(1 - g)\tau_0)},\quad C_2 = \frac{F_0 \xi (1 + 1.5\xi)}{2} - 2C_1, \tag{6.24}$$

where

$$w(\xi, \tau_0) = 1 + 1.5\xi + (1 - 1.5\xi)\exp(-\tau_0/\xi). \tag{6.25}$$

Therefore, we can derive instead of Eqs. (6.13)–(6.14):

$$\eta I_1'(\vartheta) + I_1(\vartheta) - \frac{F_0}{4}(1 - 3\xi^2)\exp(-\tau/\xi) - 3gC_1\eta + 3(1 - g)C_1\tau - C_2 = 0 \tag{6.26}$$

$$\eta I_2'(\vartheta) + I_2(\vartheta) - \frac{3S_0 g v}{4} = 0, \tag{6.27}$$

where $v = \sqrt{(1 - \xi^2)(1 - \eta^2)}$ and we assume that $h_2(\tau) = 0$. This approximation is valid because $h_2(\tau)$ is determined by $I_2(\vartheta)$, which is small at large values of τ because radiation is azimuthally independent in that scenario. Equations (6.26)–(6.27) can be solved analytically, providing results for $I_1(\vartheta)$ and $I_2(\vartheta)$ and, therefore, $I(\vartheta, \varphi)$ (see Eq. (6.12)) for the linear phase function given by Eq. (6.1). Alternatively, we can write the following equation for the intensity $I(\vartheta, \varphi)$:

$$\eta I'(\vartheta) + I(\vartheta) - 3gC_1\eta + 3(1 - g)C_1\tau - C_2 + \frac{F_0}{4}(1 - 3\xi^2 + 3gv\cos(\varphi - \varphi_0))e^{-\tau/\xi} = 0, \tag{6.28}$$

which can be solved analytically. However, our intention is to generalize Eq. (6.28) for the case of arbitrary phase functions. This can be done in an approximate way, assuming that the only change of Eq. (6.28) is the subtraction of the single-scattering term for the single-scattering phase function (Eq. (6.1)) by the corresponding term for the arbitrary phase function. Then it follows that:

$$\eta I'(\vartheta) + I(\vartheta) = B(\tau), \tag{6.29}$$

where the source function can be presented by the following analytical equation:

$$B(\tau) = B_{ss}(\tau) + B_{ms}(\tau), \tag{6.30}$$

where

$$B_{ss}(\tau) = F_0 p(\theta)\exp(-\tau/\xi)/4 \tag{6.31}$$

and

$$B_{ms}(\tau) = -3gC_1\eta + 3(1 - g)C_1\tau - C_2 + \frac{F_0}{4}(3\xi^2 + 3g\eta\xi)e^{-\tau/\xi} \tag{6.32}$$

are the source functions for the single- and multiple-scattered light, respectively, at $\omega_0 = 1$. The solution of Eq. (6.29) with boundary Eq. (6.6) can be represented as follows:

$$I^{\downarrow}(\tau) = \frac{1}{\eta}\int_0^{\tau} B(\tau')\exp((\tau' - \tau)/\eta)d\tau', \eta > 0, \tag{6.33}$$

$$I^{\uparrow}(\tau) = \frac{1}{\eta}\int_{\tau_0}^{\tau} B(\tau')\exp((\tau' - \tau)/\eta)d\tau', \eta < 0, \tag{6.34}$$

where arrows represent downward and upward propagating radiation, respectively. Equations (6.33)–(6.34) are general solutions of the RTE valid for arbitrary source functions. These equations have been used to derive the main formulae of the single-scattering approximation assuming that $B(\tau) = \omega_0 B_{ss}(\tau)$ (see Section 3.1). Now we shall use these equations to derive the approximate solutions of the RTE valid at arbitrary optical thickness. Let us substitute Eqs. (6.30)–(6.32) in Eqs. (6.33)–(6.34). Then we derive the following approximate solutions of the integro-differential RTE:

$$\begin{aligned} I^{\downarrow}(\tau) &= (3C_1\eta + C_2)\left(e^{\frac{\tau-\tau_0}{\eta}} - e^{-\frac{\tau_0}{\eta}}\right) - 3(1-g)C_1\tau e^{\frac{\tau-\tau_0}{\eta}} \\ &+ \frac{F_0\xi e^{-\frac{\tau_0}{\eta}}}{4(\eta - \xi)}\left(p(\theta_t) - 3\xi^2 - 3g\eta\xi\right)\left(e^{\left[\frac{1}{\eta}+\frac{1}{\xi}\right]\tau} - 1\right), \end{aligned} \tag{6.35}$$

$$\begin{aligned} I^{\uparrow}(\tau) &= (3C_1\eta - C_2)\left(e^{-\frac{\tau_0}{\eta}} - e^{-\frac{\tau}{\eta}}\right) - 3(1-g)C_1\left(\tau e^{-\frac{\tau}{\eta}} + \tau_0 e^{-\frac{\tau_0}{\eta}}\right) \\ &+ \frac{F_0\xi}{4(\eta + \xi)}\left(p(\theta_r) - 3\xi^2 + 3g\eta\xi\right)\left(e^{-\left[\frac{1}{\eta}+\frac{1}{\xi}\right]\tau} - e^{-\left[\frac{1}{\eta}+\frac{1}{\xi}\right]\tau_0}\right), \end{aligned} \tag{6.36}$$

where the integration constants are given by Eq. (6.24). We have substituted η by $-\eta$ in Eq. (6.36). Therefore, η is positive in both equations. It should be pointed out that:

$$\begin{aligned} \cos\theta_t &= \xi\eta + \sqrt{(1-\xi^2)(1-\eta^2)}\cos(\varphi - \varphi_0), \\ \cos\theta_r &= -\xi\eta + ss_0\cos(\varphi - \varphi_0). \end{aligned} \tag{6.37}$$

These analytical equations are very important because they make it possible to estimate light intensities at any level inside the turbid media and also at their boundaries.

Let us consider the reflection and transmission functions:

$$\begin{aligned} R &= \frac{\pi I^{\uparrow}(0)}{\xi F_0}, \\ T &= \frac{\pi I^{\downarrow}(\tau_0)}{\xi F_0}. \end{aligned} \tag{6.38}$$

Then the aforementioned equations are simplified:

$$\begin{aligned} R &= 1 - \frac{w(\tau_0,\xi)w(\tau_0,\eta)}{4 + 3(1-g)\tau_0} \\ &+ \frac{1}{4(\eta+\xi)}(p(\theta) + 3(1+g)\eta\xi - 2(\xi+\eta))(1 - e^{-m\tau_0}), \end{aligned} \tag{6.39}$$

$$T = \frac{w(\tau_0,\xi)w(\tau_0,\eta)}{4+3(1-g)\tau_0} - \frac{1}{2}\left[e^{-\frac{\tau_0}{\eta}} - e^{-\frac{\tau_0}{\xi}}\right] + \frac{1}{4(\eta-\xi)}(p(\theta) - 3(1+g)\eta\xi)\left[e^{-\frac{\tau_0}{\eta}} - e^{-\frac{\tau_0}{\xi}}\right] \tag{6.40}$$

where $m = \xi^{-1} + \eta^{-1}$.

The accuracy of Eqs. (6.39)–(6.40) decreases with the value of the asymmetry parameter g. Their range of validity can be increased if one uses a truncation approximation. In the framework of this approximation, the real phase function is substituted by the sum of the phase function extended in the forward direction ($p_f(\theta)$) and another smooth phase function ($p_b(\theta)$).

Examples of the reflection function computed using the Sobolev approximation (Eq. (6.39)) are shown in Figures 6.1 and 6.2 for Rayleigh scattering. The relative error of this approximation as a function of the optical thickness is shown in Figure 6.3. Formulas similar to Eqs. (6.39)–(6.40) except with accounting for light absorption are given by Sobolev (1956).

The errors of Eq. (6.39) which account for the truncation approximation for aerosol phase functions have been studied by Katsev et al. (2010). It has been found that the errors are below 10% for the nadir observation conditions and typical atmospheric aerosol concentrations.

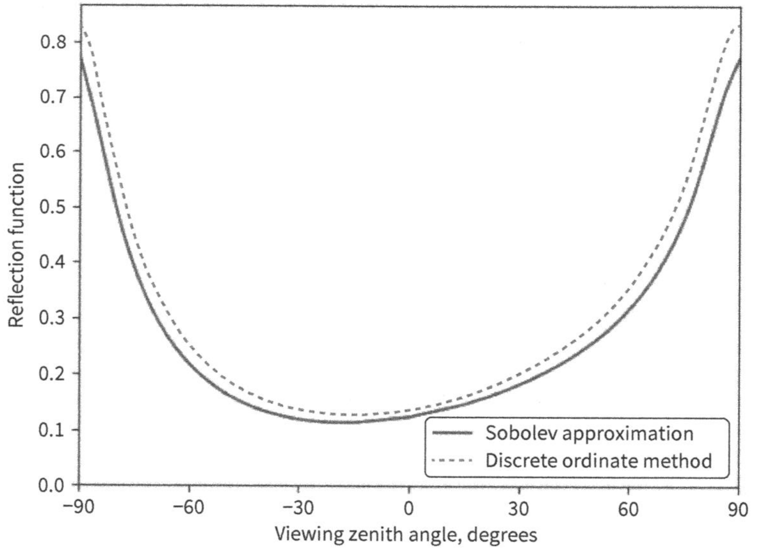

Figure 6.1 The dependence of the reflection function on the viewing zenith angle (Eq. (6.39)) and the discrete ordinate method for Rayleigh scattering. The optical thickness of the layer is 0.3, the single-scattering albedo is 1.0, and the incident zenith angle is 60°.

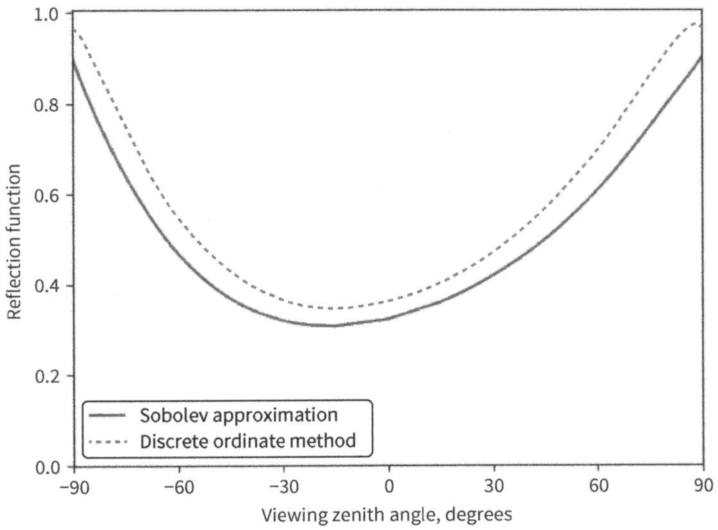

Figure 6.2 Same as Figure 6.1, but for an optical thickness of 1.0.

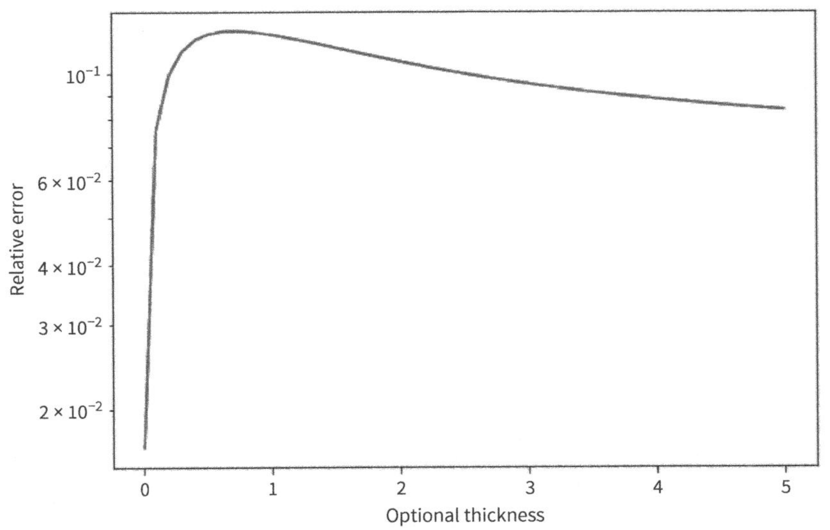

Figure 6.3 The relative error of the Sobolev approximation for computing the reflection function, as a function of the optical thickness, for Rayleigh scattering. The incident zenith angle is 60°, while the viewing zenith angle is 0°.

6.2 Two-stream Approximation

The discrete ordinates method can be used to numerically solve the RTE. The details of this method can be found in Efremenko et al. (2017b). A practical guide for numerical implementation of the discrete ordinate method is given in Korkin et al. (2022). The following discussion considers particular cases of the discrete ordinates method, viz., the two-stream and four-stream approximations.

As shown in Section 2.9, the azimuthal dependence can be factored out of the RTE by expanding the intensity in a Fourier cosine series. This results in the RTE splitting up into $2N$ independent equations, one for each azimuthal intensity component, where $2N$ is the order of the Legendre polynomial expansion of the phase function.

$$\mu \frac{dI^m(\tau,\mu)}{d\tau} = I^m(\tau,\mu) - J^m(\tau,\mu), \qquad m = 0, 1, \ldots, 2N-1, \tag{6.41}$$

where $J^m(\tau,\mu)$ is the source term and is given by:

$$\begin{aligned} J^m(\tau,\mu) = & \int_{-1}^{1} \Pi^m(\mu,\mu')\, I^m(\tau,\mu')\, d\mu' \\ & + \frac{F_0}{2\pi}(2-\delta_{m0})\,\Pi^m(\mu,-\mu_0)\exp(-\tau/\mu_0) \end{aligned} \tag{6.42}$$

The first term on the right-hand side of Eq. (6.42) is the multiple-scatter source term, and the second term is the single-scatter source term. F_0 is the incoming solar flux. The auxiliary quantities Π^m are defined in terms of normalized associated Legendre polynomials $Y_l^m(\mu)$ and phase function moments β_l:

$$\Pi^m(\mu,\mu') = \frac{\omega}{2}\sum_{l=m}^{2N-1} \beta_l Y_l^m(\mu)\, Y_l^m(\mu'). \tag{6.43}$$

To obtain discrete ordinates solutions, we replace the multiple-scatter integral in Eq. (6.42) with a summation using two Gauss–Legendre quadratures defined separately in each polar angle half-space. Each quadrature has N points (streams).

The two-stream approximation is a seminal concept in the field of radiative transfer that simplifies the complex interactions of light within a medium by reducing the infinite dimensions of angular variability to just two directions. This approach has been instrumental in atmospheric science, astrophysics, and climate modeling, among other areas, by providing a manageable yet insightful framework for understanding radiative processes.

The genesis of the two-stream approximation can be traced back to the early 20th century, with significant contributions made by several scientists, including Arthur Schuster (Schuster, 1905) and Karl Schwarzschild (Schwarzschild, 1906). Schuster introduced the idea of approximating the scattering of light in a slab of material by considering only two directions, while Schwarzschild developed equations for radiative equilibrium considering the absorption and emission of radiation in stellar atmospheres. The two-stream approximation was truly advanced in the 1940s and 1950s into a robust analytical tool in the work of Chandrasekhar (1950).

The two-stream approximation simplifies the RTE by considering radiation traveling in only two directions: upward and downward. This reduction dramatically simplifies the mathematical treatment of radiative transfer, making it possible to derive analytical solutions for otherwise intractable problems. The approximation is particularly useful in optically thick media, where detailed angular resolution of the radiation field is less critical than understanding the overall balance of fluxes.

In the two-stream approximation, the azimuthal expansion of the radiation field is restricted to the first two terms; the phase function Legendre polynomial expansion also has the same restriction. Equation (6.41) can then be expressed as:

$$\mu \frac{dI^m(\tau,\mu)}{d\tau} = I^m(\tau,\mu) - \int_{-1}^{1} \Pi^m(\mu,\mu')I^m(\tau,\mu')d\mu' \\ -F_0 Q^m(\mu)\exp(-\tau/\mu_0), \qquad m = 0, 1, \tag{6.44}$$

where

$$Q^m(\mu) = \frac{1}{2\pi}(2-\delta_{m0})\Pi^m(\mu,-\mu_0). \tag{6.45}$$

The integral in Eq. (6.44) is approximated by a summation using double Gauss–Legendre quadrature, with one stream value, $\bar{\mu}$, in each half-space.

We now omit the superscript m for the sake of clarity, and define the following quantities:

$$I^{\pm} = I(\tau, \pm\bar{\mu}), \tag{6.46}$$

$$M^{\pm} = I^{+} \pm I^{-}. \tag{6.47}$$

Equation (6.44) can then be written as:

$$\frac{dM^{\pm}}{d\tau} = -(\zeta \mp \eta)\, M^{\mp} - \frac{F_0}{\bar{\mu}}\,(Q^{+} \mp Q^{-})\exp(-\tau/\mu_0), \tag{6.48}$$

where

$$Q^{\pm} = Q(\pm\bar{\mu}), \tag{6.49}$$

$$\zeta = \frac{1}{\bar{\mu}}\,(\Pi^{+} - 1), \tag{6.50}$$

$$\eta = \frac{1}{\bar{\mu}}\Pi^{-}, \tag{6.51}$$

$$\Pi^{\pm} = \Pi(\bar{\mu}, \pm\bar{\mu}). \tag{6.52}$$

The general solution to Eq. (6.48) can be written as:

$$M^{\pm} = \tilde{M}^{\pm} + W^{\pm}F_0\exp(-\tau/\mu_0), \tag{6.53}$$

where $\tilde{M}^{\pm}$ and $W^{\pm}F_0\exp(-\tau/\mu_0)$ are, respectively, the homogeneous and particular solutions to Eq. (6.48).

The homogeneous solutions take the form:

$$M^{\pm} = Y^{\pm}\exp(-k\tau), \tag{6.54}$$

and we obtain the following eigenvalue problem:

$$\begin{bmatrix} -k & d \\ s & -k \end{bmatrix}\begin{bmatrix} Y^{+} \\ Y^{-} \end{bmatrix} = 0, \tag{6.55}$$

where

$$s = \zeta + \eta, \tag{6.56}$$

$$d = \zeta - \eta. \tag{6.57}$$

The eigenvalues of the system of Eq. (6.55) can be easily computed:

$$k^{\pm} = \pm\sqrt{sd}. \tag{6.58}$$

The eigenvectors can be obtained by substituting for the eigenvalues into Eq. (6.55):

$$Y^{+} = 1, \tag{6.59}$$

$$Y^{-} = \pm s/k, \tag{6.60}$$

where $k = \sqrt{sd}$, and the positive and negative values in Eq. (6.60) are for the positive and negative eigenvalues, respectively.

The particular solutions can be obtained by substituting Eqs. (6.53)–(6.60) into Eq. (6.48):

$$W^{+} = \frac{1}{\mu_0^{-2} - sd}\left[\mu_0^{-1}Z^{-} + dZ^{+}\right], \tag{6.61}$$

$$W^{-} = \frac{1}{\mu_0^{-2} - sd}\left[sZ^{-} + \mu_0^{-1}Z^{+}\right], \tag{6.62}$$

where

$$Z^{\pm} = \frac{Q^{+} \pm Q^{-}}{\bar{\mu}}. \tag{6.63}$$

The general solution to Eq. (6.48) can then be written as:

$$M^{+} = Ae^{-k\tau} + Be^{k\tau} + W^{+}F_0e^{-\tau/\mu_0}, \tag{6.64}$$

$$M^{-} = A\frac{s}{k}e^{-k\tau} - B\frac{s}{k}e^{k\tau} + W^{-}F_0e^{-\tau/\mu_0}, \tag{6.65}$$

where A and B are the constants of integration.

If we now stratify the atmosphere into N layers, with τ_{p-1} and τ_p the optical depths at the top and bottom of layer p, respectively ($\tau_0 = 0$), we can write the following expressions for $I^{\pm}$:

$$\begin{aligned} I^{\pm} &= A_p\left(\frac{1 \pm s_p/k_p}{2}\right)e^{-k_p(\tau-\tau_{p-1})} + B_p\left(\frac{1 \mp s_p/k_p}{2}\right)e^{-k_p(\tau_p-\tau)} \\ &+ \left(\frac{W_p^{+} \pm W_p^{-}}{2}\right)F_0e^{-\tau/\mu_0}, (p = 1, \ldots, N), \end{aligned} \tag{6.66}$$

where A_p and B_p are the constants of integration in layer p. Exponential arguments in Eq. (6.66) have been written as optical depth differences to safeguard the numerical stability of the solution (van Oss and Spurr, 2002).

The integration constants follow a set of three boundary conditions:

1. Zero diffuse downwelling field at the upper boundary;
2. Radiance continuity at intermediate layer boundaries; and
3. Surface reflection function at the lower boundary.

Solving the boundary value problem is the only numerical part of the two-stream solution. The boundary condition at the top of the atmosphere reads:

$$A_1\left(\frac{1-s_1/k_1}{2}\right)+B_1\left(\frac{1+s_1/k_1}{2}\right)e^{-k_1\tau_1}=-\left(\frac{W_1^+-W_1^-}{2}\right)F_0. \tag{6.67}$$

The layer interface conditions are as follows:

$$\begin{aligned}&A_{p-1}\left(\frac{1\pm s_{p-1}/k_{p-1}}{2}\right)e^{-k_{p-1}(\tau_{p-1}-\tau_{p-2})}+B_{p-1}\left(\frac{1\mp s_{p-1}/k_{p-1}}{2}\right)\\&-A_p\left(\frac{1\pm s_p/k_p}{2}\right)-B_p\left(\frac{1\mp s_p/k_p}{2}\right)e^{-k_p(\tau_p-\tau_{p-1})}=\\&\left(\frac{\{W_p^+\pm W_p^-\}-\{W_{p-1}^+\pm W_{p-1}^-\}}{2}\right)F_0e^{-\tau_{p-1}/\mu_0},(p=2,\ldots,N)\end{aligned} \tag{6.68}$$

For the boundary condition at the bottom of the atmosphere, we have:

$$\begin{aligned}&A_N\left(\frac{\overset{\frown}{Y}{}_N^{+}+\overset{\frown}{Y}{}_N^{-}}{2}\right)e^{-k_N(\tau_N-\tau_{N-1})}+B_N\left(\frac{\overset{\frown}{Y}{}_N^{+}-\overset{\frown}{Y}{}_N^{-}}{2}\right)=\\&-\left(\frac{\hat{W}_N^++\hat{W}_N^-}{2}\right)F_0e^{-\tau_N/\mu_0}+\frac{\mu_0}{\pi}F_0e^{-\tau_N/\mu_0}\rho^m(\mu_0,\bar{\mu}),\end{aligned} \tag{6.69}$$

where $\rho^m(\mu_0,\bar{\mu})$ is the m-th Fourier component of the bidirectional reflection distribution function (BRDF) for incident angle $\arccos(\mu_0)$ and outgoing angle $\arccos(\bar{\mu})$, and:

$$\overset{\frown}{Y}{}_N^{+}=1-(1+\delta_{0m})\bar{\mu}\rho^m(\bar{\mu},\bar{\mu}), \tag{6.70}$$

$$\overset{\frown}{Y}{}_N^{-}=\frac{s_N}{k_N}\left[1+(1+\delta_{0m})\bar{\mu}\rho^m(\bar{\mu},\bar{\mu})\right], \tag{6.71}$$

$$\hat{W}_N^+=W_N^+\left[1-(1+\delta_{0m})\bar{\mu}\rho^m(\bar{\mu},\bar{\mu})\right], \tag{6.72}$$

$$\hat{W}_N^-=W_N^-\left[1+(1+\delta_{0m})\bar{\mu}\rho^m(\bar{\mu},\bar{\mu})\right]. \tag{6.73}$$

For a Lambertian surface with albedo a:

$$\rho^m(\mu_0,\bar{\mu})=\rho^m(\bar{\mu},\bar{\mu})=\delta_{0m}a. \tag{6.74}$$

It should be noted that the boundary value problem is the only part of the two-stream solution requiring nonanalytic techniques. However, the linear system of equations for A_p and B_p has a pentadiagonal form; therefore, specialized pentadiagonal solvers can be used instead of traditional matrix inversion routines.

The discrete ordinates solution gives the radiance at the Gaussian quadrature angle. However, the viewing direction does not typically coincide with this angle. Therefore, post processing is required to obtain results at the required viewing angle. This can be done by substituting the discrete ordinates solution for the quadrature angle into the RTE, and integrating the latter. In an inhomogeneous atmosphere, the source terms must be integrated on a layer-by-layer basis. We obtain the following for the upwelling radiance at the top of the atmosphere:

$$I^m(0,\mu) = I^m(\tau_N,\mu)e^{-\tau_N/\mu} + \sum_{p=1}^{N}\int_{\tau_{p-1}}^{\tau_p}\left[\Pi^m(\mu,\bar{\mu})I^+ + \Pi^m(\mu,-\bar{\mu})I^- + F_0Q^m(\mu)e^{-\tau/\mu_0}\right]\frac{e^{-\tau/\mu}}{\mu}d\tau\ (m=0,1), \tag{6.75}$$

The optical depth integration can be performed analytically to obtain:

$$I^m(0,\mu) = I^m(\tau_N,\mu)e^{-\tau_N/\mu} + \sum_{p=1}^{N} e^{-\tau_{p-1}/\mu}\begin{bmatrix} A_p\{X_p^+(\mu)+Z_p^-(\mu)\}E_p^+(\mu) \\ +B_p\{X_p^-(\mu)+Z_p^+(\mu)\}E_p^-(\mu) \\ +F_0\{V_p^+(\mu)+V_p^-(\mu)\}E_p^0(\mu) \\ +F_0Q^m(\mu)E_p^0(\mu)\end{bmatrix}(m=0,1), \tag{6.76}$$

where

$$X_p^+(\mu) = \frac{1}{2}\Pi^m(\mu,\bar{\mu})\left(1+s_p/k_p\right)(p=1,\ldots,N), \tag{6.77}$$

$$X_p^-(\mu) = \frac{1}{2}\Pi^m(\mu,\bar{\mu})\left(1-s_p/k_p\right)(p=1,\ldots,N), \tag{6.78}$$

$$Z_p^+(\mu) = \frac{1}{2}\Pi^m(\mu,-\bar{\mu})\left(1+s_p/k_p\right)(p=1,\ldots,N), \tag{6.79}$$

$$Z_p^-(\mu) = \frac{1}{2}\Pi^m(\mu,-\bar{\mu})\left(1-s_p/k_p\right)(p=1,\ldots,N), \tag{6.80}$$

$$V_p^+(\mu) = \frac{1}{2}\Pi^m(\mu,-\bar{\mu})\left(W_p^+ + W_p^-\right)(p=1,\ldots,N), \tag{6.81}$$

$$V_p^-(\mu) = \frac{1}{2}\Pi^m(\mu,-\bar{\mu})\left(W_p^+ - W_p^-\right)(p=1,\ldots,N), \tag{6.82}$$

$$E_p^+(\mu) = \frac{1-e^{-\Delta\tau_p/\mu}e^{-k_p\Delta\tau_p}}{1+\mu k_p}\ (p=1,\ldots,N), \tag{6.83}$$

$$E_p^-(\mu) = \frac{e^{-k_p \Delta\tau_p} - e^{-\Delta\tau_p/\mu}}{1 - \mu k_p} \quad (p = 1, \ldots, N), \tag{6.84}$$

$$E_p^0(\mu) = \frac{e^{-\tau_{p-1}/\mu_0} - e^{-\Delta\tau_p/\mu} e^{-\tau_p/\mu_0}}{1 + \mu/\mu_0} \quad (p = 1, \ldots, N). \tag{6.85}$$

It remains to find the boundary source term $I^m(\tau_N, \mu)$, which is the m-th Fourier component of the upwelling radiation at the surface in the viewing direction. This is given by

$$\begin{aligned} I^m(\tau_N, \mu) = (1 + \delta_{m0})\bar{\mu}\rho^m(\bar{\mu}, \mu) & \left[\begin{array}{l} \left\{ \begin{array}{l} A_N \left(\frac{1 - s_N/k_N}{2} \right) e^{-k_N \Delta\tau_N} \\ +B_N \left(\frac{1 + s_N/k_N}{2} \right) \end{array} \right\} \\ + \left(\frac{W_N^+ - W_N^-}{2} \right) F_0 e^{-\tau_N/\mu_0} \end{array} \right] \\ & + F_0 \frac{\mu_0}{\pi} e^{-\tau_N/\mu_0} \rho^m(\mu_0, \mu) \, (m = 0, 1). \end{aligned} \tag{6.86}$$

Using a similar approach, results can be obtained for upwelling or downwelling radiance at desired viewing angles anywhere in the atmosphere. The two-stream single-scatter contributions are based on severely truncated phase function information, and in practical applications, it makes sense to replace these terms with accurate single-scatter radiative transfer calculations using the full phase function. The two-stream model is then simply used to compute the multiple-scattering contribution. When used this way, the only change would be to ignore the last term on the right-hand side of Eq. (6.86).

Typically, two-stream calculations are used for the computation of fluxes and mean intensities, which involve only the azimuth independent component ($m = 0$). These quantities can be obtained from Eq. (6.86) by setting $m = 0$. The reader is also referred to Liou (2002a) for the derivation of these quantities.

6.3 Four-stream Approximation

The four-stream approximation in radiative transfer is an extension of the two-stream approximation, designed to provide a more detailed and accurate representation of the angular distribution of radiation within a medium. While the two-stream approximation simplifies the RTE by considering only upward and downward fluxes, the four-stream method further divides the radiation field into four distinct directions. This division typically includes two upward and two downward directions, often angled to better capture the anisotropic nature of scattering and absorption phenomena in various media.

Equation (6.41) can then be expressed as:

$$\begin{aligned} \mu \frac{dI^m(\tau, \mu)}{d\tau} &= I^m(\tau, \mu) - \int_{-1}^{1} \Pi^m(\mu, \mu') I^m(\tau, \mu') \, d\mu' \\ &\quad - F_0 Q^m(\mu) \exp(-\tau/\mu_0) \\ &\quad (m = 0, \ldots, 3), \end{aligned} \tag{6.87}$$

where

$$Q^m(\mu) = \frac{1}{2\pi}(2 - \delta_{m0})\,\Pi^m(\mu, -\mu_0). \tag{6.88}$$

The integral in Eq. (6.87) is approximated by a summation using double Gauss–Legendre quadrature, with abscissae μ_i and weights a_i, $i = 1, 2$ in the positive half-space. The values in the negative half-space satisfy the following relations:

$$\mu_{-i} = -\mu_i, \quad i = 1, 2, \tag{6.89}$$

$$a_{-i} = a_i, \quad i = 1, 2. \tag{6.90}$$

We now omit the superscript m for the sake of clarity and define the following quantities:

$$I_i^{\pm} = I(\tau, \pm\mu_i), \quad i = 1, 2, \tag{6.91}$$

$$M_i^{\pm} = I_i^{+} \pm I_i^{-}, \quad i = 1, 2. \tag{6.92}$$

Equation (6.87) can then be written as:

$$\frac{dM_i^{\pm}}{d\tau} = -\sum_{j=1}^{2}(\zeta_{ij} \mp \eta_{ij})\,M_j^{\mp} - \frac{F_0}{\mu_i}(Q_i^{+} \mp Q_i^{-})\,exp(-\tau/\mu_0), \quad i = 1, 2, \tag{6.93}$$

where

$$Q_i^{\pm} = Q(\pm\mu_i), \quad i = 1, 2, \tag{6.94}$$

$$\zeta_{ij} = \frac{1}{\mu_i}(a_j\Pi_{ij}^{+} - \delta_{ij}), \quad i = 1, 2, \quad j = 1, 2, \tag{6.95}$$

$$\eta_{ij} = \frac{1}{\mu_i}a_j\Pi_{ij}^{-}, \quad i = 1, 2, \quad j = 1, 2, \tag{6.96}$$

$$\Pi_{ij}^{\pm} = \Pi(\mu_i, \pm\mu_j), \quad i = 1, 2, \quad j = 1, 2, \tag{6.97}$$

and δ_{ij} refers to the delta function.

Equation (6.93) can be simplified to obtain a single equation for M_i^{-}:

$$\frac{d^2M_i^{-}}{d\tau^2} = \sum_{j=1}^{2}\Gamma_{ij}M_j^{-} + d_iF_0\exp(-\tau/\mu_0), \quad i = 1, 2, \tag{6.98}$$

where

$$\Gamma_{ij} = \sum_{k=1}^{2}(\zeta_{ik} + \eta_{ik})(\zeta_{kj} - \eta_{kj}), \quad i = 1, 2, \quad j = 1, 2, \tag{6.99}$$

$$d_i = \frac{1}{\mu_i\mu_0}(Q_i^{+} + Q_i^{-}) + \frac{1}{\mu_i}\sum_{j=1}^{2}(\zeta_{ij} + \eta_{ij})(Q_i^{+} - Q_i^{-}), \quad i = 1, 2. \tag{6.100}$$

The general solution to Eq. (6.98) can be written as:

$$M_i^- = \tilde{M}_i^- + W_i^- F_0 \exp(-\tau/\mu_0), \quad i = 1, 2, \tag{6.101}$$

where $\tilde{M}_i^-$ and $W_i^- F_0 \exp(-\tau/\mu_0)$, $i = 1, 2$, are, respectively, the homogeneous and particular solutions to Eq. (6.98).

The homogeneous solutions take the form:

$$\tilde{M}_i^- = Y_i^- \exp(-k\tau), \quad i = 1, 2, \tag{6.102}$$

and we obtain the following eigenproblem:

$$\sum_{j=1}^{2} \Gamma_{ij} Y_j^- = \gamma Y_i^-, \quad i = 1, 2, \tag{6.103}$$

where $\gamma = \mathrm{k}^2$. Since Γ is a Hermitian matrix, all the eigenvalues are real. The eigenvectors and eigenvalues of the system of Eq. (6.103), denoted $\{Y_{ij}^-, \gamma_j\}$, $(i = 1,\ 2)$, $(j = 1,\ 2)$, can be computed as follows:

$$\gamma_1 = \frac{1}{2}(\Gamma_{11} + \Gamma_{22}) + \frac{1}{2}\sqrt{(\Gamma_{11} - \Gamma_{22})^2 + 4\Gamma_{21}\Gamma_{12}}, \tag{6.104}$$

$$\gamma_2 = \frac{1}{2}(\Gamma_{11} + \Gamma_{22}) - \frac{1}{2}\sqrt{(\Gamma_{11} - \Gamma_{22})^2 + 4\Gamma_{21}\Gamma_{12}}, \tag{6.105}$$

$$Y_{11}^- = 1, \tag{6.106}$$

$$Y_{22}^- = 1, \tag{6.107}$$

$$Y_{21}^- = \frac{\Gamma_{21}}{\gamma_1 - \Gamma_{22}}, \tag{6.108}$$

$$Y_{12}^- = \frac{\Gamma_{12}}{\gamma_2 - \Gamma_{11}}. \tag{6.109}$$

The particular solutions can be obtained by substituting Eqs. (6.101)–(6.109) into Eq. (6.98):

$$W_1^- = \frac{\Gamma_{12} d_2 + (\lambda^2 - \Gamma_{22})\, d_1}{\Psi}, \tag{6.110}$$

$$W_2^- = \frac{\Gamma_{21} d_1 + (\lambda^2 - \Gamma_{11})\, d_2}{\Psi}, \tag{6.111}$$

where

$$\Psi = \lambda^4 - (\Gamma_{11} + \Gamma_{22})\lambda^2 + \Gamma_{11}\Gamma_{22} - \Gamma_{12}\Gamma_{21}. \tag{6.112}$$

The general solution to Eq. (6.98) can then be written as:

$$M_i^- = \sum_{j=1}^{2} \left\{ L_j^+ Y_{ij}^- e^{-k_j \tau} + L_j^- Y_{ij}^- e^{+k_j \tau} \right\} + W_i^- F_0 e^{-\tau/\mu_0}, \quad i = 1, 2, \tag{6.113}$$

where L_j^+ and L_j^- are the constants of integration, and $k_j = \sqrt{\gamma_j}$, (j = 1, 2).

From Eq. (6.93), we can calculate M_i^+:

$$M_i^+ = \sum_{j=1}^{2} \left\{ L_j^+ Y_{ij}^+ e^{-k_j \tau} - L_j^- Y_{ij}^+ e^{+k_j \tau} \right\} + W_i^+ F_0 e^{-\tau/\mu_0}, \quad i = 1, 2, \tag{6.114}$$

where

$$Y_{ij}^+ = \frac{1}{k_j} \sum_{k=1}^{2} (\zeta_{ik} - \eta_{ik}) Y_{kj}^-, \quad i = 1, 2, \quad j = 1, 2, \tag{6.115}$$

$$W_i^+ = \mu_0 \sum_{j=1}^{2} (\zeta_{ij} - \eta_{ij}) W_j^- + \frac{\mu_0}{\mu_i} (Q_i^+ - Q_i^-), \quad i = 1, 2. \tag{6.116}$$

If we now stratify the atmosphere into N layers, with τ_{p-1} and τ_p the optical depths at the top and bottom of layer p, respectively ($\tau_0 = 0$), we can write the following expressions for $I_i^\mp$, ($i = 1, 2$):

$$\begin{aligned} I_i^\mp = \sum_{j=1}^{2} \left\{ L_{j,p}^+ \left(\frac{Y_{ij,p}^+ \pm Y_{ij,p}^-}{2} \right) e^{-k_{j,p}(\tau - \tau_{p-1})} - L_{j,p}^- \left(\frac{Y_{ij,p}^+ \mp Y_{ij,p}^-}{2} \right) e^{-k_{j,p}(\tau_p - \tau)} \right\} \\ + \left(\frac{W_{i,p}^+ \pm W_{i,p}^-}{2} \right) F_0 e^{-\tau/\mu_0}, \quad i = 1, 2, \quad p = 1, \ldots, N. \end{aligned} \tag{6.117}$$

where $L_{j,p}^+$ and $L_{j,p}^-$ are the constants of integration in layer p. Exponential arguments in Eq. (6.117) have been written as optical depth differences to safeguard the numerical stability of the solution (van Oss and Spurr, 2002).

The integration constants follow from a set of three boundary conditions:

1. Zero diffuse downwelling field at the upper boundary;
2. Radiance continuity at intermediate layer boundaries; and
3. Surface reflection function at the lower boundary. Solving the boundary value problem is the only numerical part of the four-stream solution.

The boundary condition at the top of the atmosphere reads:

$$\begin{aligned} \sum_{j=1}^{2} \left\{ L_{j,1}^+ \left(\frac{Y_{ij,1}^+ - Y_{ij,1}^-}{2} \right) - L_{j,1}^- \left(\frac{Y_{ij,1}^+ + Y_{ij,1}^-}{2} \right) e^{-k_{j,1}\tau_1} \right\} = \\ - \left(\frac{W_{i,1}^+ - W_{i,1}^-}{2} \right) F_0, \quad i = 1, 2. \end{aligned} \tag{6.118}$$

The layer interface conditions are as follows:

$$\begin{aligned} \sum_{j=1}^{2} \left\{ L_{j,p-1}^+ \left(\frac{Y_{ij,p-1}^+ \pm Y_{ij,p-1}^-}{2} \right) e^{-k_{j,p-1}(\tau_{p-1} - \tau_{p-2})} - L_{j,p-1}^- \left(\frac{Y_{ij,p-1}^+ \mp Y_{ij,p-1}^-}{2} \right) \right. \\ \left. - L_{j,p}^+ \left(\frac{Y_{ij,p}^+ \pm Y_{ij,p}^-}{2} \right) + L_{j,p}^- \left(\frac{Y_{ij,p}^+ \mp Y_{ij,p}^-}{2} \right) e^{-k_{j,p}(\tau_p - \tau_{p-1})} \right\} = \\ \left\{ \frac{\left(W_{i,p}^+ \pm W_{i,p}^- \right) - \left(W_{i,p-1}^+ \pm W_{i,p-1}^- \right)}{2} \right\} F_0 e^{-\tau_{p-1}/\mu_0}, \quad p = 2, \ldots, N. \end{aligned} \tag{6.119}$$

For the boundary condition at the bottom of the atmosphere, we have:

$$\sum_{j=1}^{2}\left\{L_{j,N}^{+}\left(\frac{\hat{Y}_{ij,N}^{+}+\hat{Y}_{ij,N}^{-}}{2}\right)e^{-k_{j,N}(\tau_N-\tau_{N-1})}-L_{j,N}^{-}\left(\frac{\hat{Y}_{ij,N}^{+}-\hat{Y}_{ij,N}^{-}}{2}\right)\right\}= \\ -\left(\frac{\hat{W}_{i,N}^{+}+\hat{W}_{i,N}^{-}}{2}\right)F_0e^{-\frac{\tau_N}{\mu_0}}+\frac{\mu_0}{\pi}F_0e^{-\frac{\tau_N}{\mu_0}}\rho^m(\mu_0,\mu_i),\quad i=1,2, \tag{6.120}$$

where $\rho^{\mathrm{m}}(\mu_0,\overline{\mu})$ is the m-th Fourier component of the BRDF for incident angle $\arccos(\mu_0)$ and outgoing angle $\arccos(\mu_{\mathrm{i}})$, and:

$$\hat{\mathrm{Y}}_{\mathrm{ij,N}}^{+}=\mathrm{Y}_{\mathrm{ij,N}}^{+}-(1+\delta_{0\mathrm{m}})\sum_{\mathrm{k}=1}^{2}\mathrm{a_k}\mu_{\mathrm{k}}\mathrm{Y}_{\mathrm{kj,N}}^{+}\rho^{\mathrm{m}}(\mu_{\mathrm{k}},\mu_{\mathrm{i}})(\mathrm{i}=1,\ 2),\ (\mathrm{j}=1,\ 2) \tag{6.121}$$

$$\hat{\mathrm{Y}}_{\mathrm{ij,N}}^{-}=\mathrm{Y}_{\mathrm{ij,N}}^{-}+(1+\delta_{0\mathrm{m}})\sum_{\mathrm{k}=1}^{2}\mathrm{a_k}\mu_{\mathrm{k}}\mathrm{Y}_{\mathrm{kj,N}}^{-}\rho^{\mathrm{m}}(\mu_{\mathrm{k}},\mu_{\mathrm{i}})(\mathrm{i}=1,\ 2),\ (\mathrm{j}=1,\ 2) \tag{6.122}$$

$$\hat{\mathrm{W}}_{\mathrm{i,N}}^{+}=\mathrm{W}_{\mathrm{i,N}}^{+}-(1+\delta_{0\mathrm{m}})\sum_{\mathrm{k}=1}^{2}\mathrm{a_k}\mu_{\mathrm{k}}\mathrm{W}_{\mathrm{k,N}}^{+}\rho^{\mathrm{m}}(\mu_{\mathrm{k}},\mu_{\mathrm{i}})(\mathrm{i}=1,\ 2) \tag{6.123}$$

$$\hat{\mathrm{W}}_{\mathrm{i,N}}^{-}=\mathrm{W}_{\mathrm{i,N}}^{-}+(1+\delta_{0\mathrm{m}})\sum_{\mathrm{k}=1}^{2}\mathrm{a_k}\mu_{\mathrm{k}}\mathrm{W}_{\mathrm{k,N}}^{-}\rho^{\mathrm{m}}(\mu_{\mathrm{k}},\mu_{\mathrm{i}})(\mathrm{i}=1,\ 2) \tag{6.124}$$

For a Lambertian surface with albedo a:

$$\rho^m(\mu_0,\mu_i)=\rho^m(\mu_k,\mu_i)=\delta_{0m}a,\quad i=1,\ 2,\quad k=1,\ 2. \tag{6.125}$$

The linear system of equations for $L_{j,p}^{\pm}$, $(j=1,2)$ is sparse since the boundary value problem matrix only contains nonzero terms in a band along the diagonal. This can be taken advantage of in the numerical solution.

The discrete ordinates solution gives the radiance at the Gaussian quadrature angles. However, the viewing direction does not typically coincide with one of these angles. Therefore, post processing is required to obtain results at the required viewing angle. This can be done by substituting the discrete ordinates solution for the quadrature angles into the RTE, and integrating the latter. In an inhomogeneous atmosphere, the source terms must be integrated on a layer-by-layer basis. We obtain the following for the upwelling radiance at the top of the atmosphere:

$$I^m(0,\mu)=I^m(\tau_N,\mu)e^{-\tau_N/\mu} \\ +\sum_{p=1}^{N}\int_{\tau_{p-1}}^{\tau_p}\left[\left(\sum_{i=1}^{2}a_i\{\Pi^m(\mu,\mu_i)I_i^{+}+\Pi^m(\mu,-\mu_i)I_i^{-}\}\right)+\right. \\ \left.+F_0Q^m(\mu)e^{-\tau/\mu_0}\right]\frac{e^{-\tau/\mu}}{\mu}d\tau,\quad m=0,\ldots,3. \tag{6.126}$$

The optical depth integration can be performed analytically to obtain:

$$\begin{aligned}
I^m(0,\mu) &= I^m(\tau_N,\mu)\,e^{-\tau_N/\mu} \\
&+ \sum_{p=1}^{N} e^{-\tau_{p-1}/\mu} \sum_{j=1}^{2} \left[L_{j,p}^{+} \left\{ X_{j,p}^{+}(\mu) + Z_{j,p}^{-}(\mu) \right\} E_{j,p}^{+}(\mu) \right. \\
&\left. - L_{j,p}^{-} \left\{ X_{j,p}^{-}(\mu) + Z_{j,p}^{+}(\mu) \right\} E_{j,p}^{-}(\mu) \right] + F_0 \left\{ V_p^{+}(\mu) + V_p^{-}(\mu) \right\} E_p^{0}(\mu) \\
&+ F_0 Q^m(\mu)\, E_p^0(\mu)\,(m = 0,\ 1,\ 2,\ 3)
\end{aligned} \tag{6.127}$$

where

$$X_{j,p}^{+}(\mu) = \frac{1}{2} \sum_{i=1}^{2} a_i \Pi^m(\mu,\mu_i) \left(Y_{ij,p}^{+} + Y_{ij,p}^{-} \right) (j = 1,\ 2), (p = 1,\ \ldots,\ N) \tag{6.128}$$

$$X_{j,p}^{-}(\mu) = \frac{1}{2} \sum_{i=1}^{2} a_i \Pi^m(\mu,\mu_i) \left(Y_{ij,p}^{+} - Y_{ij,p}^{-} \right) (j = 1,\ 2), (p = 1,\ \ldots,\ N) \tag{6.129}$$

$$Z_{j,p}^{+}(\mu) = \frac{1}{2} \sum_{i=1}^{2} a_i \Pi^m(\mu,-\mu_i) \left(Y_{ij,p}^{+} + Y_{ij,p}^{-} \right) (j = 1,\ 2), (p = 1,\ \ldots,\ N) \tag{6.130}$$

$$Z_{j,p}^{-}(\mu) = \frac{1}{2} \sum_{i=1}^{2} a_i \Pi^m(\mu,-\mu_i) \left(Y_{ij,p}^{+} - Y_{ij,p}^{-} \right) (j = 1,\ 2), (p = 1,\ \ldots,\ N) \tag{6.131}$$

$$V_p^{+}(\mu) = \frac{1}{2} \sum_{i=1}^{2} a_i \Pi^m(\mu,\mu_i) \left(W_{i,p}^{+} + W_{i,p}^{-} \right) (p = 1,\ \ldots,\ N) \tag{6.132}$$

$$V_p^{-}(\mu) = \frac{1}{2} \sum_{i=1}^{2} a_i \Pi^m(\mu,-\mu_i) \left(W_{i,p}^{+} - W_{i,p}^{-} \right) (p = 1,\ \ldots,\ N) \tag{6.133}$$

$$E_{j,p}^{+}(\mu) = \frac{1 - e^{-\Delta\tau_p/\mu} e^{-k_{j,p}\Delta\tau_p}}{1 + \mu k_{j,p}} (j = 1,\ 2), (p = 1,\ \ldots,\ N) \tag{6.134}$$

$$E_{j,p}^{-}(\mu) = \frac{e^{-k_{j,p}\Delta\tau_p} - e^{-\Delta\tau_p/\mu}}{1 - \mu k_{j,p}} (j = 1,\ 2), (p = 1,\ \ldots,\ N) \tag{6.135}$$

$$E_p^{0}(\mu) = \frac{e^{-\tau_{p-1}/\mu_0} - e^{-\Delta\tau_p/\mu} e^{-\tau_p/\mu_0}}{1 + \mu/\mu_0} (p = 1,\ \ldots,\ N) \tag{6.136}$$

It remains to find the boundary source term $I^m(\tau_N, \mu)$, which is the m-th Fourier component of the upwelling radiation at the surface in the viewing direction. This is given by:

$$I^m(\tau_N, \mu) = (1+\delta_{0m}) \sum_{i=1}^{2} a_i \mu_i \rho^m(\mu_i, \mu) \times \left[\sum_{j=1}^{2} \left\{ L^+_{j,N} \left(\frac{Y^+_{ij,N} - Y^-_{ij,N}}{2} \right) e^{-k_{j,N}\Delta\tau_N} - L^-_{j,N} \left(\frac{Y^+_{ij,N} + Y^-_{ij,N}}{2} \right) \right\} + \left(\frac{W^+_{i,N} - W^-_{i,N}}{2} \right) F_0 e^{-\tau_N/\mu_0} \right] + F_0 \frac{\mu_0}{\pi} e^{-\frac{\tau_N}{\mu_0}} \rho^m(\mu_0, \mu) \, (m = 0, 1, 2, 3). \tag{6.137}$$

Using a similar approach, results can be obtained for upwelling or downwelling radiance at desired viewing angles anywhere in the atmosphere.

The two- and four-stream methods can be generalized to cases with an arbitrary number of discrete ordinates. In this generalized case, the overall approach remains the same. However, the eigenvalue problem can no longer be solved analytically. To compute the eigenvalues and eigenvectors, specialized solvers can be employed, such as the ASYMTX solver,[1] which are specifically designed for efficient handling of such decompositions (see Stamnes et al. (1988) and references therein).

6.4 The Spherical Harmonics Method

The spherical harmonics method was proposed by Jeans to solve the radiative transfer problem in stars. Further description of the method, as it applies to radiative transfer, was given by Kourganoff (1952). The idea of the method consists of expanding the radiance and the single-scattering phase function into a finite series of spherical harmonics. In this way, the initial RTE turns into a system of coupled partial differential equations for the expansion coefficients of the radiance.

Let us consider the RTE for the azimuth harmonics of the radiance (see Eqs. (2.66), (2.67), and (2.75)):

$$\mu \frac{\mathrm{dI}_m}{\mathrm{d}\tau}(\tau, \mu) = -\mathrm{I}_m(\tau, \mu) + \frac{\omega_0}{2} \int_{-1}^{1} \mathrm{I}_m(\tau, \mu') p_m(\mu, \mu') \, d\mu'. \tag{6.138}$$

We expand the radiance I_m into a series of associated Legendre polynomials:

$$\mathrm{I}_m(\tau, \mu) = \sum_{n=m}^{\infty} (2n+1) \frac{(n-m)!}{(n+m)!} \mathrm{I}^m_n(\tau) P^m_n(\mu) P^m_n(\mu'). \tag{6.139}$$

The same procedure is applied for the single-scattering phase function:

$$p_m(\mu, \mu') = \sum_{n=m}^{\infty} (2n+1) \frac{(n-m)!}{(n+m)!} \chi_n P^m_n(\mu) P^m_n(\mu'). \tag{6.140}$$

Note that in the small-angle approximation, in the left-hand side of Eq. (6.138), there was μ_0 instead of μ. And so, all the terms of the equation had the same expansion kernels. Now

1 ASYMTX is a modification of the LAPACK routine that deals only with real roots.

the problem appears with the differential term of Eq. (6.138). To expand $\mu I_m(\tau,\mu)$ in terms of Legendre polynomials, we make use of the following property:

$$\mu \bar{P}_n^m(\mu) = \frac{(n+1)\bar{P}_{n+1}^m(\mu)}{\sqrt{2n+1}\sqrt{2n+3}}\sqrt{\frac{(l+m+1)}{(l-m+1)}} + \frac{n\bar{P}_{n-1}^m(\mu)}{\sqrt{2n+1}\sqrt{2n-1}}\sqrt{\frac{(l-m)}{(l+m)}}, \tag{6.141}$$

where

$$\bar{P}_n^m(\mu) = \sqrt{(2n+1)\frac{(n-m)!}{(n+m)!}}P_n^m(\mu) \tag{6.142}$$

are normalized Legendre polynomials. Thus, the left-hand side of Eq. (6.138) turns into

$$\mu\frac{dI_m(\tau,\mu)}{d\tau} = \sum_{n=m}^{\infty}\left\{\frac{(n+1)\bar{P}_{n+1}^m(\mu)}{\sqrt{2n+1}\sqrt{2n+3}}\sqrt{\frac{(l+m+1)}{(l-m+1)}} + \frac{n\bar{P}_{n-1}^m(\mu)}{\sqrt{2n+1}\sqrt{2n-1}}\sqrt{\frac{(l-m)}{(l+m)}}\right\}\frac{dI_n^m(\tau)}{d\tau} \tag{6.143}$$

Then, by shifting the indices of summation, we obtain

$$\mu\frac{dI_m(\tau,\mu)}{dz} = \sum_{n=m}^{\infty}\left\{\frac{n\bar{P}_n^m(\mu)}{\sqrt{2n-1}\sqrt{2n+1}}\sqrt{\frac{(n+m)}{(n-m)}}\frac{dI_{n-1}^m(\tau)}{d\tau} + \frac{(n+1)\bar{P}_n^m(\mu)}{\sqrt{2n+1}\sqrt{2n+3}}\sqrt{\frac{(n-m+1)}{(n+m+1)}}\frac{dI_{n+1}^m(\tau)}{d\tau}\right\}. \tag{6.144}$$

Substituting Eqs. (6.139), (6.140), and (6.144) into Eq. (6.138) leads to an infinite system of equations with respect to I_n^m. In order to make the system closed, higher-order expansion coefficients can be neglected. In the so-called P_1 approximation (also referred to as the diffuse approximation), the expansion terms with $n = 0$ and $n = 1$ are retained (detailed derivations can be found in Case and Zweifel (1967)). The P_1 approximation is useful for studies of radiative transfer in optically thick turbid media.

6.5 Phase Function Truncation Methods

In radiative transfer modeling, particularly when using the discrete ordinate method, the phase function is commonly expanded in terms of Legendre polynomials. This expansion efficiently represents the angular scattering properties of a medium. However, phase functions describing scattering by aerosol and cloud particles often exhibit a sharp forward peak due to diffraction effects, requiring a large number of expansion coefficients for accurate representation. In some cases, thousands of expansion terms may be necessary, leading to computationally expensive simulations. To alleviate this issue, various phase function truncation techniques have been introduced, aimed at approximating the original function with fewer terms while maintaining numerical accuracy in radiative transfer solutions.

The challenge of handling highly anisotropic scattering phase functions became apparent in the early developments of multiple scattering theories in atmospheric and astrophysical contexts. Early studies in radiative transfer, such as those by Chandrasekhar (1950), highlighted the difficulty of resolving forward peaks while keeping calculations tractable. In astrophysics, the problem of approximating scattering properties efficiently was also encountered in modeling stellar atmospheres, as noted in the works of Sobolev (1975). The need for computationally efficient techniques led to the development of truncation strategies that simplify phase functions while preserving their primary scattering characteristics.

One of the earliest and most widely used approximations, first introduced by Potter (1970), involves replacing the forward diffraction peak with a delta function while maintaining a smoothed residual phase function:

$$p(\cos\Theta) \approx 2f\delta(1-\cos\Theta) + \tilde{p}(\cos\Theta), \tag{6.145}$$

where f represents the fraction of total scattering concentrated in the forward peak, and $\tilde{p}(\cos\Theta)$ is a modified phase function that remains unchanged outside a defined forward cone. The parameter defining this forward-scattering region, $\tilde{\mu}$, is often chosen based on empirical estimates; for instance, Hansen (1969) suggested a value of approximately 0.94 for cloud phase functions.

Building on this concept, the so-called δ-S method (Mitrescu and Stephens, 2004) was developed, wherein the forward-scattering lobe is replaced with a uniform value inside the truncation region. This approach offers a balance between computational efficiency and accuracy, making it particularly useful in atmospheric remote sensing applications.

A more refined method was introduced with the δ-M approximation by Wiscombe and Evans (1977). In this method, the forward peak is approximated by a delta function, and the remaining phase function is scaled accordingly:

$$p(\cos\Theta) = 2f\delta(1-\cos\Theta) + (1-f)\tilde{p}(\cos\Theta). \tag{6.146}$$

This formulation ensures that the integral of the phase function over all angles remains conserved. Expanding Eq. (6.146) in Legendre polynomials,

$$2f\delta(1-\cos\Theta) = \sum_{k=0}^{\infty}(2k+1)fP_k(\cos\Theta), \tag{6.147}$$

$$\tilde{p}(\cos\Theta) = \sum_{k=0}^{N_{\mathrm{tr}}}(2k+1)\tilde{\chi}_k P_k(\cos\Theta), \tag{6.148}$$

one finds that the expansion coefficients are related by:

$$\tilde{\chi}_k = \frac{\chi_k - f}{1-f}, \quad k = 0, \ldots, N_{\mathrm{tr}}. \tag{6.149}$$

This transformation modifies the RTE by introducing scaled optical parameters:

$$\tilde{\sigma}_{\mathrm{ext}} = \sigma_{\mathrm{ext}} - \sigma_{\mathrm{sca}} f, \tag{6.150}$$

$$\tilde{\sigma}_{\mathrm{sca}} = \sigma_{\mathrm{sca}} \frac{1-f}{1-f\omega}, \tag{6.151}$$

$$\tau' = \tau(1 - \omega f), \tag{6.152}$$

$$\omega' = \omega \frac{1-f}{1-\omega f}. \tag{6.153}$$

Here ω is the single-scattering albedo. The delta-M method introduces errors due to the nature of the truncation applied to the forward peak of the phase function. When we substitute Eqs. (6.147), (6.148), and (6.149) into Eq. (6.146), the approximate scattering phase function takes the form:

$$p(\cos\Theta) = (1-f)\sum_{k=0}^{M}(2k+1)\widetilde{\chi}_k P_k(\cos\Theta) + \sum_{k=M+1}^{\infty}(2k+1)fP_k(\cos\Theta). \tag{6.154}$$

We can note that in Eq. (6.154) the first M terms preserve the original Legendre expansion coefficients. This means that the approximation maintains accuracy up to the truncation order M. However, the error due to truncation is embedded in the higher-order terms ($k > M$), where the expansion coefficients remain artificially constant at f instead of following the original phase function moments.

The consequence of this truncation is that the phase function representation remains accurate for forward-scattering angles but introduces deviations at larger angles, particularly in the backscattering regime ($\cos\Theta < 0$). The truncation essentially removes the variability of higher-order terms, leading to systematic discrepancies in the phase function at oblique angles.

Since the residual error is predominantly concentrated in the neglected higher-order terms, it is particularly significant when dealing with strongly peaked phase functions, such as those associated with Mie scattering. The errors introduced by the delta-M truncation may lead to an underestimation of multiple scattering effects and affect the accuracy of radiative transfer simulations, particularly in optically thick media.

This limitation motivated the development of the δ-fit method (Hu et al., 2000), which minimizes the residual between $\tilde{p}$ and p by solving a least-squares optimization problem:

$$r_{\delta\text{-fit}} = \sum_{i=1}^{q}\left(\frac{\tilde{p}(\mu_i) - p(\mu_i)}{p(\mu_i)}\right)^2 w_i, \tag{6.155}$$

where w_i are weights assigned to selected scattering angles outside the forward peak region. This method significantly improves accuracy in backscattering while maintaining computational efficiency.

More recent refinements of truncation techniques include the work of Lin et al. (2018), who replaced the delta function with a forward-peaked function of finite width:

$$2f\delta(1-\cos\Theta) \approx \sum_{k=0}^{\infty}(2k+1)c\exp\left(-\frac{k^2}{2\sigma^2}\right)fP_k(\cos\Theta), \tag{6.156}$$

where c and σ are determined by moment-matching criteria.

Although phase function truncation methods provide significant computational benefits, they invariably introduce some degree of error. The extent of this error depends on the choice of truncation parameters and the application-specific tolerances. Figures 6.4 and 6.5 illustrate examples of truncated phase functions compared to their full-resolution counterparts.

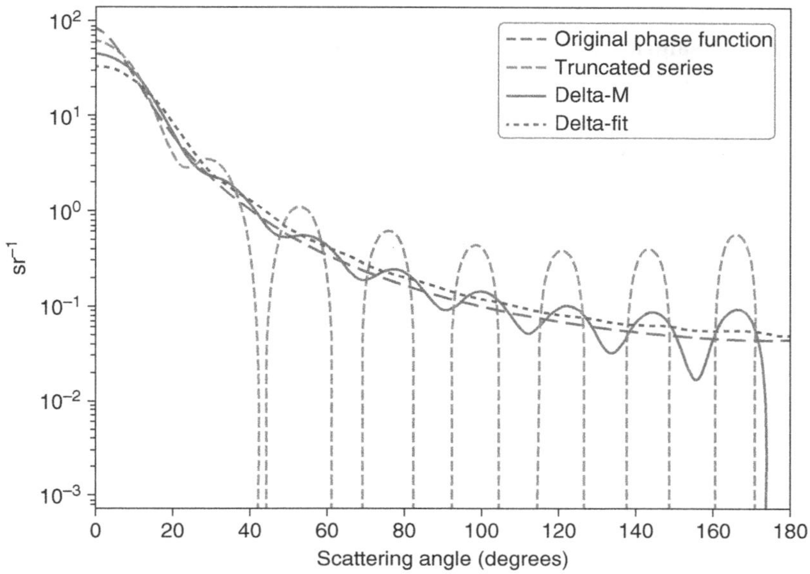

Figure 6.4 Comparison of phase function truncation methods for a Henyey–Greenstein phase function with asymmetry parameter 0.85.

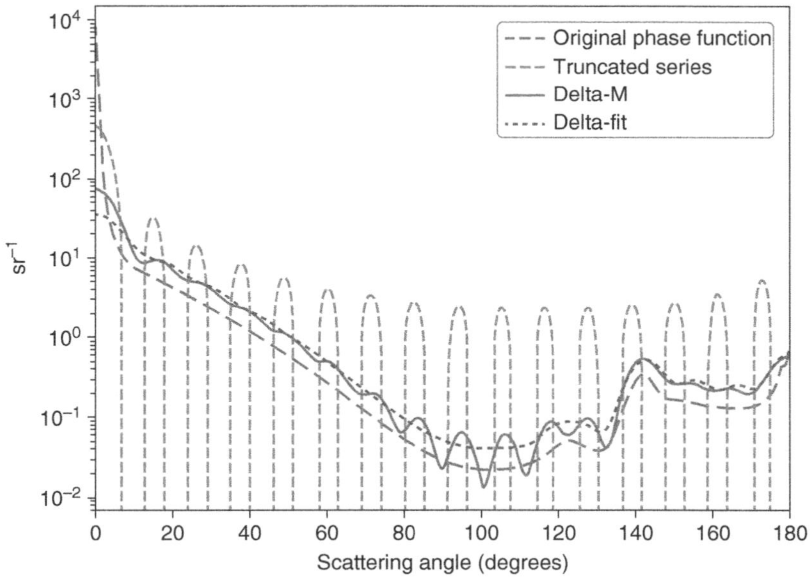

Figure 6.5 Same as Figure 6.4, but for a water cloud phase function.

Overall, truncation techniques have played a crucial role in advancing radiative transfer modeling, allowing for efficient solutions without excessively compromising accuracy. Their continued refinement remains an active area of research, particularly as high-resolution remote sensing applications demand ever-greater accuracy in scattering representations.

7

Radiative Transfer in Gaseous Absorption Bands

7.1 *k*-Distribution and Correlated-*k* Methods

Atmospheric radiative transfer is affected by various processes: gas absorption, molecular (Rayleigh) scattering, and scattering and absorption by clouds and aerosols. The most accurate way to perform radiative transfer calculations is to do them in a monochromatic fashion, where cross-sections are calculated for each absorption line in the spectral region of interest. The problem is that such line-by-line (LBL) calculations are computationally expensive and not operationally viable. For this reason, a range of approximations have been used to accelerate the computations. This section discusses these methods.

In exploring the effect of absorption lines on stellar radiative equilibrium, Ambartzumian (1936) came up with the idea that transmission within a spectral interval depends only on the frequency distribution of absorption coefficient (k) strengths in the interval, not on their LBL variation. The k-distribution method relies on grouping spectral intervals according to absorption coefficient strength. Very few points are required in the k-distribution compared to that needed for LBL calculations, leading to high computational efficiency. Arking and Grossman (1972) employed this methodology to analyze the radiative equilibrium temperature profile in planetary atmospheres.

To extend this concept to inhomogeneous atmospheres, Lacis et al. (1979) and Chou and Arking (1980) separated variables and wrote the absorption coefficient as a product of two factors: one depending only on the wavenumber and the other only on temperature and pressure. Lacis et al. (1979) were evaluating the impact on cirrus clouds on surface temperature, while Chou and Arking (1980) were attempting to compute atmospheric cooling rates in water vapor absorption channels in the infrared spectral region. In the troposphere and lower stratosphere, where molecular collisions are the main cause of line broadening, a Lorentzian line shape can be employed. For typical atmospheric conditions, the absorption cross-section k at a pressure p and temperature T can be calculated as follows:

$$k(p, T) = k(p_\mathrm{r}, T_\mathrm{r})\frac{p}{p_\mathrm{r}}f(T, T_\mathrm{r}), \tag{7.1}$$

where p_r is a reference pressure and T_r is a reference temperature. f is a function of temperature but not pressure. Clearly, the right-hand side is composed of terms that are a function either of the wavenumber or of pressure and temperature (but not both). The term $k(p_r, T_r)$ can be computed offline for the relevant spectral regions, thereby leading to significant

Analytical Methods in Radiative Transfer, First Edition. Alexander Kokhanovsky, Vijay Natraj and Dmitry Efremenko.

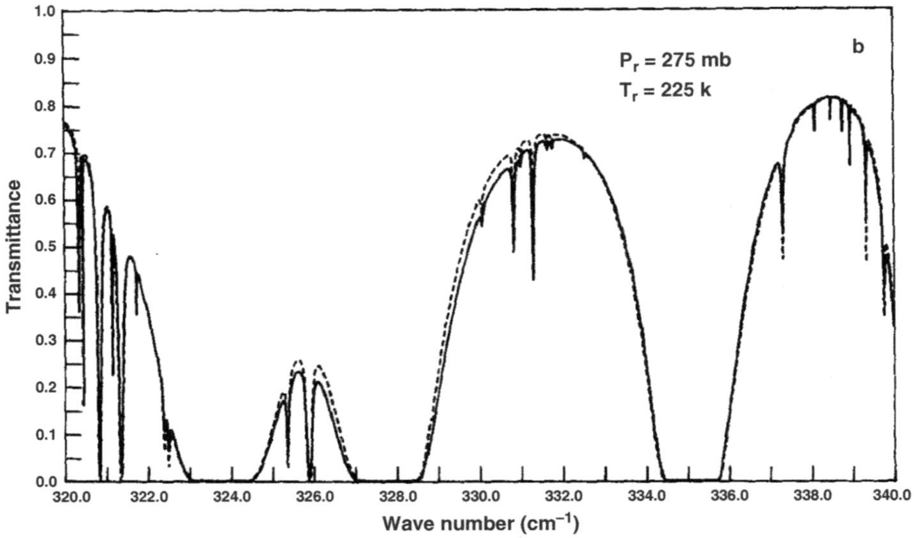

Figure 7.1 Transmittance comparison between LBL and far-wing scaling calculations. The solid curve represents LBL calculations and the dashed curve represents far-wing scaling calculations. *Source*: Used with permission from Chou and Arking (1980). © American Meteorological Society.

speed improvements. Figure 7.1 shows a comparison of transmittances computed using the LBL approach (solid curve) with those calculated using the scaling described in Eq. 7.1 (dashed curve).

Goody et al. (1989) and Lacis and Oinas (1991) introduced the correlated-k technique, which is an extension of the k-distribution method for vertically inhomogeneous atmospheres. They proceeded as follows. First, define a function $f(k)$, which is the frequency distribution of absorption coefficients in a spectral region. The fraction of the spectral range for which the absorption coefficient is less than k, or equivalently, the cumulative distribution function $g(k)$, can be expressed as follows:

$$g(k) = \int_0^k f(k')\,dk'. \tag{7.2}$$

The k-distribution function $k(g)$ is then the inverse of the cumulative frequency distribution. Normalizing the frequency distribution to unity ensures that g is in the range [0, 1]. There is a unique solution for g when the atmosphere is homogeneous, as it is a monotonic function of k in this scenario, making it possible to move seamlessly from frequency to g space. The situation is more complicated for inhomogeneous atmospheres, where g is a function of altitude, making the inversion non-unique. Goody et al. (1989) made the simplifying (correlated-k) assumption that the k-distribution at one vertical atmospheric level is also valid for other pressure levels. In other words, the spectral variation of absorption coefficient strengths remains the same at all altitudes. This assumption is accurate for weak and strong absorptions. Goody et al. (1989) and Lacis and Oinas (1991) demonstrated that this technique was especially useful for the calculation of atmospheric heating and cooling rates.

Buchwitz et al. (2000) extended the correlated-k approach to enable its usage for spectral windows containing overlapping line and/or continuum absorbers. The k-binning method was developed by Boesche et al. (2008). While the conventional k-distribution approach requires calculations to be done for each instrument channel, k-binning performs simulations of the entire absorption band, thereby obviating the need for prior knowledge of the sensor response function. Cao et al. (2011) optimized the number of k-distribution bins and related parameters, including the corresponding absorption coefficients and quadrature weights, with special procedures in place to avoid ill-conditioning.

7.2 Exponential Sum Fitting of Transmittances

In order to further improve on the correlated-k approach, Hunt and Grant (1969) optimized the number of k values using a procedure called exponential sum fitting of transmissions (ESFT). Wiscombe and Evans (1977) independently describe the theoretical basis for this technique. The basic idea is to fit transmission functions with exponential sums for calculating spectrally integrated radiative fluxes, especially when both line absorption and scattering are important. The spectral mean transmission $T(u)$ for absorber amount u is calculated as follows:

$$T(u) \cong E(u) = \sum_{i=1}^{M} w_i \exp(-k_i u), \tag{7.3}$$

where E indicates the sum of exponentials, and M is the number of monochromatic wavelengths. The weights w_i satisfy a normalization constraint:

$$\sum_{i=1}^{M} w_i = 1. \tag{7.4}$$

The weights are determined using the following two-step procedure:

1. Calculate a set $u_n = n\Delta u,\ n = 0, \ldots, N$ of evenly spaced values between 0 and u. $\Delta u = u/N$.
2. Perform least squares minimization for the residual R:

$$R = \sum_{n=0}^{N} [T(u_n) - E(u_n)]^2. \tag{7.5}$$

The ESFT technique was employed for analyzing advanced very-high-resolution radiometer measurements by Kratz (1995). Kratz et al. (1998) applied the technique to calculate radiative forcing due to minor trace gas absorption in the 800–1380 cm^{-1} infrared spectral region. Chou et al. (1993) demonstrated that dividing the water vapor and CO_2 absorption spectrum in the infrared makes flux calculations more efficient. Mano (1995) permitted unequal wavelength spacing and introduced a nonlinear least squares procedure for calculating the weights, thereby overcoming some of the limitations of the traditional ESFT approach. Armbruster and Fischer (1996) improved the accuracy of the ESFT technique by accounting for the pressure and temperature dependence of absorption lines.

Figure 7.2a shows the cooling rate in the 800–1380 cm^{-1} spectral window due to the minor trace gases (CO_2, N_2O, CH_4, CFC-11, CFC-12, and HCFC-22) for an atmosphere that

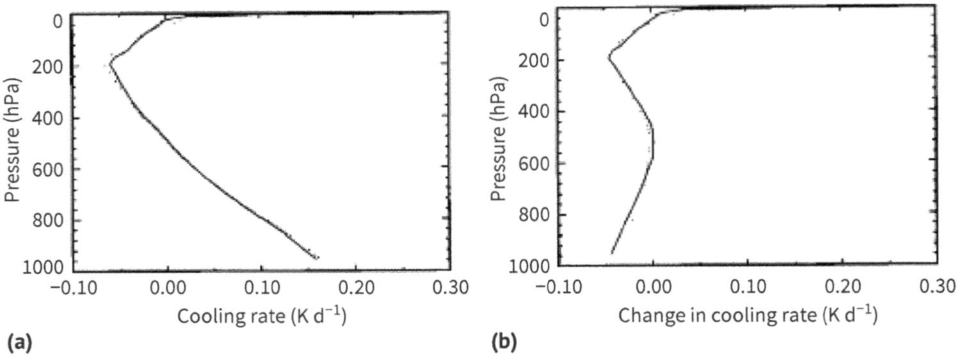

Figure 7.2 Cooling rate profiles for the 800–1380 cm^{-1} spectral range calculated using LBL (solid curve) and ESFT (dashed curve) calculations. (a) Cooling rates resulting from the introduction of the minor trace gases into an atmosphere where no other absorbers are considered. (b) Change in cooling rates resulting from the introduction of the minor trace gases into an atmosphere already containing H_2O and O_3. *Source*: Used with permission from Kratz et al. (1998). © American Geophysical Union.

has no other absorbing gases. The trace gases have a net cooling effect near the surface, a warming effect higher up in the tropopause, and then a cooling effect in the stratosphere. Figure 7.2b shows the change in the cooling rates when the minor trace gases are introduced into an atmosphere that already contains H_2O and O_3. The high concentration of water vapor in the lower troposphere reduces the cooling effect due to the minor trace gases. As the water vapor concentration reduces higher up in the atmosphere, the results look very similar to those for an atmosphere with only the minor trace gases. The dashed (ESFT) and solid (LBL) curves are indistinguishable, indicating that the ESFT technique captures the atmospheric features with high accuracy compared to the reference LBL calculations.

7.3 Spectral Mapping

The main drawback of the correlated-k technique is that it assumes that absorption coefficients with similar values at one atmospheric level remain spectrally correlated throughout the atmosphere. Therefore, while it is a good approximation for calculating broadband heating and cooling rates, it results in large errors when computing radiances. Simply increasing the number of terms for the integration in Eq. (7.5) does not necessarily lead to the correct result. West et al. (1990) realized that the correlated-k method could be viewed as a special case of a more general spectral mapping concept. Spectral mapping methods identify wavelengths that have similar optical properties throughout the atmospheric column. This method offers three advantages over the correlated-k approach. First, it enables tuning to obtain arbitrary accuracy with respect to LBL calculations. Second, it allows modeling of gas mixtures. Third, it allows for spectral dependence of the top of the atmosphere solar flux.

A numerical implementation of mapping techniques was provided by the spectral mapping atmospheric radiative transfer (SMART) model (Meadows and Crisp, 1996; Crisp, 1997). SMART

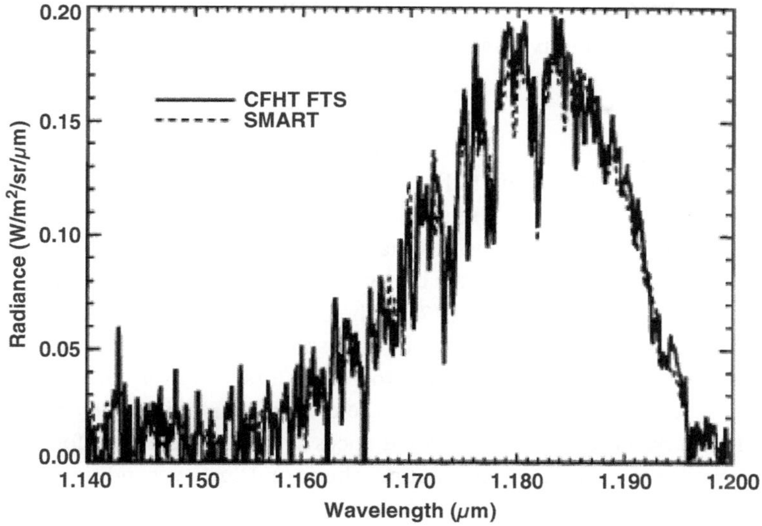

Figure 7.3 A synthetic spectrum generated using spectral mapping, compared to a high-resolution spectrum of the Venus atmosphere. *Source*: Used with permission from Meadows and Crisp (1996). © American Geophysical Union.

1. identifies monochromatic wavelengths within the desired spectral range that have similar optical properties throughout the atmospheric column,
2. groups these wavelengths into bins,
3. performs RT calculations for each bin, and
4. maps the computed bin radiances back to the monochromatic wavelength grid.

Figure 7.3 shows a sample SMART synthetic radiance spectrum, and the corresponding high-resolution spectrum of the Venus atmosphere acquired using the Canada-France-Hawaii Telescope (CFHT) Fourier transform spectrometer (FTS) (Meadows and Crisp, 1996).

7.4 Optimal Spectral Sampling

The assumption made by the correlated-k and ESFT approaches that the spectral correlation of absorption coefficients at one altitude persists at other altitudes is invalid in two scenarios:

1. for a single gas, when there are lines with different strengths or when the line strengths are highly temperature-dependent, and
2. for gas mixtures, when the relative concentrations of absorbers in the mixture vary with altitude.

Moncet et al. (2008) developed the optimum spectral sampling (OSS) model, which is a fast and accurate transmittance parameterization technique that extends the ESFT and k-distribution techniques to vertically inhomogeneous atmospheres with overlapping absorbing species. The OSS method approximates the atmospheric radiance in a given instrument channel as a weighted sum of monochromatic radiance evaluated at selected wavelengths (nodes) ν_i within the interval spanned by the instrument function:

$$T(u) \cong \sum_{i=1}^{M} w_i \exp\left[-k(\nu_i)\,u\right]. \tag{7.6}$$

Note that this is very similar to Eq. 7.5, with the difference being that spectral points are used rather than absorption cross-sections. This subtle but important change readily allows for extension to inhomogeneous atmospheres. All that would be needed in that scenario would be to employ an internal double sum over layers and gases for the terms inside the existing summation on the right-hand side of Eq. (7.6). The OSS method is equivalent to the ESFT technique in the special case of a homogeneous atmosphere with a single absorber.

Node selection is performed using the following procedure. First, for every instrument channel, N uniformly spaced wavelengths that cover the channel bandwidth are chosen. Second, a reference LBL model is used to compute radiances at these wavelengths for a set of S training profiles chosen to span the range of atmospheric and surface conditions and viewing geometries to which the OSS model will be applied. Third, the monochromatic radiances are convolved with the instrument response function to produce channel radiances, $\bar{R}_s$, for each member s of the set of training profiles. Fourth, the wavelengths ν_i and weights w_i are obtained by minimizing the root mean square (rms) difference ε_N between the true and estimated radiances.

$$\varepsilon_N = \sqrt{\frac{1}{S}\sum_{s=1}^{S}\left[\bar{R}_s - \sum_{i=1}^{N} w_i R_s(\nu_i)\right]^2}. \tag{7.7}$$

A Monte-Carlo search method is used to identify the optimal set of nodes as the smallest set of N nodes such that ε_N is less than a prescribed tolerance. The weights are normalized by requiring their sum to equal unity.

The OSS method was designed to model radiances measured by Earth-orbiting down-looking microwave and infrared radiometers, but it can be applied to any spectral domain and instrument viewing geometry, and for flux and radiance computations in emitting and scattering atmospheres. The initial investment in LBL calculations can be large because many such calculations may need to be performed to identify optimal wavelengths and associated weights.

This method was tested on simulated Atmospheric Infrared Sounder measurements. Figure 7.4 shows a comparison of analytical Jacobians (partial derivatives of radiance with respect to geophysical parameters of interest) from OSS with finite difference Jacobians calculated using the LBL radiative transfer model. OSS produces accurate temperature, water vapor, and ozone Jacobians.

7.5 Double-*k*, Linear-*k*, and Low-streams Interpolation Techniques

Duan (2005) recognized the importance of accounting for the vertical distribution of gaseous absorption in the presence of multiple scattering. The major drawback of the correlated-*k* technique is that it assumes that the ranking of absorption coefficient strengths is the same throughout the atmosphere. This assumption fails when the line absorption

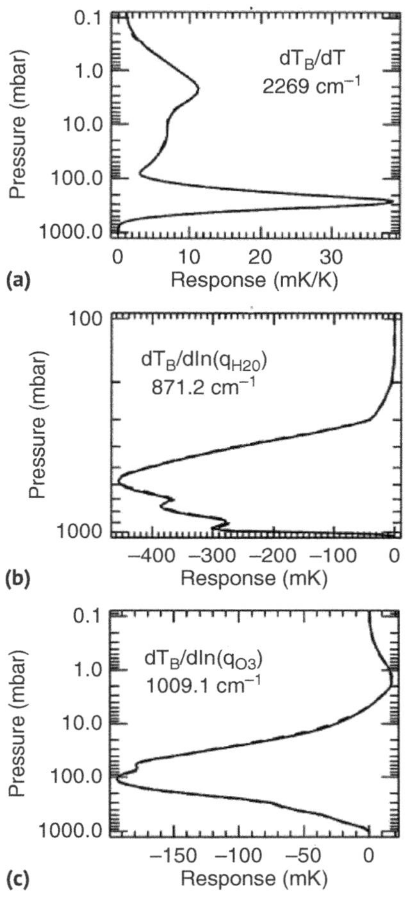

Figure 7.4 Comparison between LBL radiative transfer model (solid) and OSS (dashed) Jacobians for (a) temperature, (b) water vapor, and (c) ozone. The derivatives were converted from radiance to equivalent brightness temperature and, for (b) and (c), they are with respect to the logarithm of the mixing ratio. *Source*: Used with permission from Moncet et al. (2008). © American Meteorological Society.

profile is different even though the column optical depth is the same (see Figure 7.5 for an example of different vertical profiles of gas absorption for the same total absorption optical depth). To overcome this, Duan (2005) introduced a double-k approach. In addition to the total absorption optical thickness k, they employed the absorption optical thickness k^* from the top of the atmosphere to a "scattering" layer. The scattering layer is defined as the layer where the column scattering optical depth from the top of the atmosphere to the layer under consideration reaches half the total column scattering optical depth or unity, whichever is smaller.

Another innovation introduced by Duan (2005) was to separate the RT calculations into single and multiple scattering components. The basic concept here is that single scattering captures most of the angular variation in the intensity and is computationally inexpensive. On the other hand, multiple scattering calculations are time-intensive.

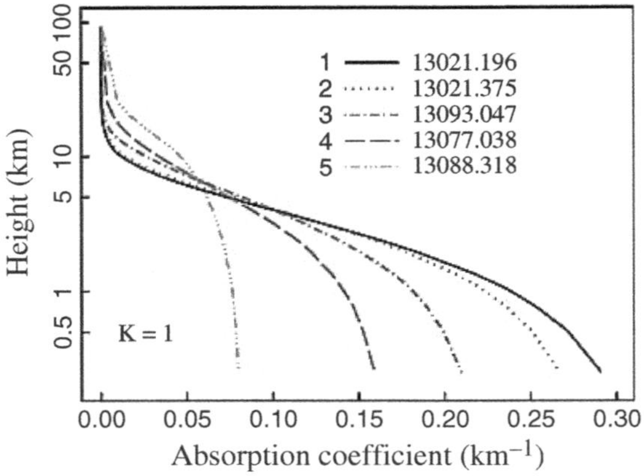

Figure 7.5 Vertical distributions of oxygen absorption at five wavenumbers with the same total absorption optical depth. *Source*: Used with permission from Duan (2005). © American Geophysical Union.

However, the multiple scattering contribution is a smooth function of absorption optical thickness. It should, therefore, be possible to evaluate the multiple scatter field efficiently. Duan (2005) computed single scattering accurately in a monochromatic fashion. The multiple scattering contribution to the radiance I^{ms} is calculated as follows:

$$I^{ms} = g(k) f_k (k^*/k), \tag{7.8}$$

where g and f_k are defined as piecewise analytical functions, with the unknown coefficients in those functions determined by fitting I^{ms} at selected k and k^* values.

Hasekamp and Butz (2008) accounted for the vertical distribution of atmospheric gaseous absorption in a different manner and extended the double-k approach to include polarization. They considered the multiple scatter component of the radiance as a function of the total absorption optical thickness and its normalized vertical distribution, where the latter is a vector that represents the relative contribution of each atmospheric layer to the total absorption optical depth. Their linear-k approach employs the following steps. First, calculations of the multiple scatter radiance component are performed for a limited number of absorption optical depths and vertical distributions. Second, for each optical depth grid point, differences in the actual vertical distribution and the distribution used in the reference calculation are accounted for by performing a linear Taylor series expansion (hence the name linear-k) around the reference distribution. Third, a second-order polynomial fit corrects for the difference between the total absorption optical thickness and the reference value. Spectral variation of atmospheric scattering and surface reflection properties are also accounted for through a linear approximation. The linear-k approach requires derivatives of the Stokes parameters with respect to the layer absorption optical thickness.

The low-streams interpolation (LSI) approach developed by O'Dell (2010) is similar to Duan (2005) in terms of separation of single- and multiple-scattering components of the radiance. There are two main differences between the two methods. First, LSI performs low- (two computational quadrature angles or "streams") and high- (large number of streams) accuracy multiple scattering calculations and reconstructs the difference between

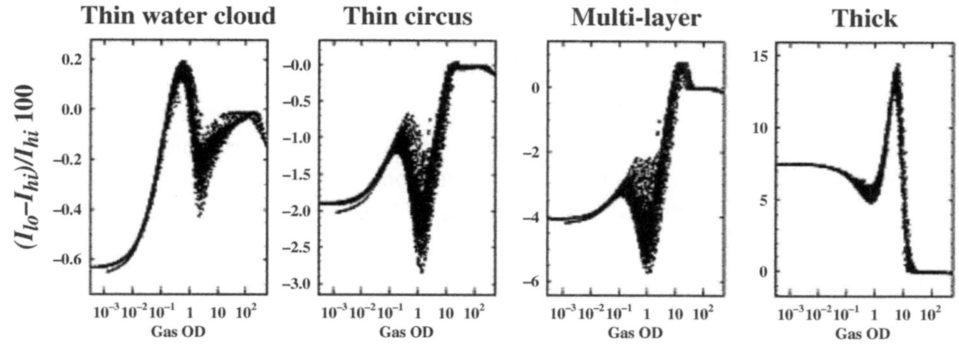

Figure 7.6 Error in Stokes parameter *I* for two-stream calculations compared to high-accuracy calculations. Errors in *I* are expressed in percent. *Source*: Used with permission from O'Dell (2010). © American Geophysical Union.

the two. The reconstruction is essentially a two-dimensional interpolation as a function of the column-integrated absorption optical depth and a parameter closely related to the quantity k^*/k in Eq. 7.8 . Second, as in Hasekamp and Butz (2008), polarization effects are accounted for.

Here, we discuss the scalar scenario. Figure 7.6 illustrates the errors in Stokes parameters I and Q for two-stream calculations relative to high-accuracy calculations, as a function of the total column gaseous absorption optical depth. The relative difference exhibits a strong correlation with the gas absorption optical depth τ_g. These results indicate that two-stream calculations could be performed at all wavelengths and the error curve reconstructed by interpolation using a small number of high-accuracy calculations. The scatter in Figure 7.6 is the result of an interplay between the vertical structure of gaseous absorption and the location of the scattering species. LSI performs a bilinear interpolation of relative radiance errors (between two-stream and high-accuracy calculations) in terms of the two variables τ_g and $\xi^{1/2}$, where ξ is defined as follows:

$$\xi = \frac{\tau_g'}{\tau_g}, \tag{7.9}$$

where τ_g' is the cumulative gaseous absorption optical depth from the top of the atmosphere down to the layer where the corresponding cumulative scattering optical depth equals a critical value. The critical value is unity or half the total scattering optical depth, whichever is smaller. ξ is very similar to the parameter k^*/k in Eq. 7.8, except that the former is the gas absorption optical depth while the latter is the absorption optical depth due to all species (including aerosols, for example).

The monochromatic value of I can be obtained as follows:

$$I = \frac{I^{\text{lo}}}{1+\varepsilon_I}. \tag{7.10}$$

where I^{lo} is the two-stream estimate of the radiance, and ε_I is the error in Stokes parameter I for the relevant wavelength computed using the bilinear interpolation described above. The rms radiance errors for the LSI technique are typically less than 0.1.

The cluster low-streams regression method described in del Águila et al. (2020) is similar to the LSI technique. Here, the low-stream (e.g. two-stream) radiative transfer model is employed on a high-resolution wavelength grid. The radiances values from these calculations are grouped into clusters. A multiple-scattering radiative transfer model is invoked for some spectral points in each cluster. Linear regression is then performed between the low-stream and multi-stream solutions. This regression model is used to restore the multi-stream solutions for the high-resolution spectral grid. In del Águila and Efremenko (2021), the cluster low-streams regression method is applied to compute the multi-stream radiances from the single-scattering solution.

7.6 Computations in a Broad Spectral Range: 400–2500 nm

The k-distribution methods have some drawbacks. First, the radiance depends not only on the absorption coefficient but also on the scattering coefficient and hence, there is no one-to-one correspondence between the k-value and the radiance value (however, it is implicitly assumed in the k-distribution method). Consequently, the k-distribution method may be inaccurate as the spectral range increases and is usually applied within a given absorption band. Second, due to the multiple-scattering impact and screening effect of a turbid medium, the impact of the absorption processes can be reduced, and the radiance dependency on k may not be purely exponential, which leads again to errors in the computations. In fact, as discussed in Fu and Liou (1992), the k-distribution method is valid only in the context of a single absorption line, periodic lines, and the strong- and weak-line limits. Finally, the result of the k-distribution method is the convolved spectrum on a coarse wavelength grid, rather than the spectrum on an initial fine grid.

Efremenko et al. (2023) proposed the following approach. For a set of N spectral points with the wavelengths λ_i, $i = 1, \ldots, N$ and set of layers $j = 1, \ldots, N_l$, we compute scattering σ_{ij} and absorption k_{ij} coefficients. The radiances $R_m = R(\lambda_{ms})$ are computed at each sth spectral point, where $m = s \cdot i$ (hereinafter, s is referred to as the sampling rate), at M spectral points in total ($M < N$). The transmittances $T_{\sigma i}$ and T_{ki} are computed as:

$$T_{\sigma i} = e^{-\sum_{j=1}^{N_l} \sigma_{ij} h_j \beta}, \tag{7.11}$$

$$T_{ki} = e^{-\sum_{j=1}^{N_l} k_{ij} h_j \beta}, \tag{7.12}$$

where

$$\beta = \frac{1}{\mu_0} + \frac{1}{\mu}, \tag{7.13}$$

h_j is the geometrical thickness of the jth layer, while μ_0 and μ are the cosines of the solar zenith angle and viewing zenith angle, respectively. Then a predictor vector p_i can be taken as:

$$p_i = \left[T_{\sigma i}, T_{ki}, \frac{1}{\lambda_i^4} \right]. \tag{7.14}$$

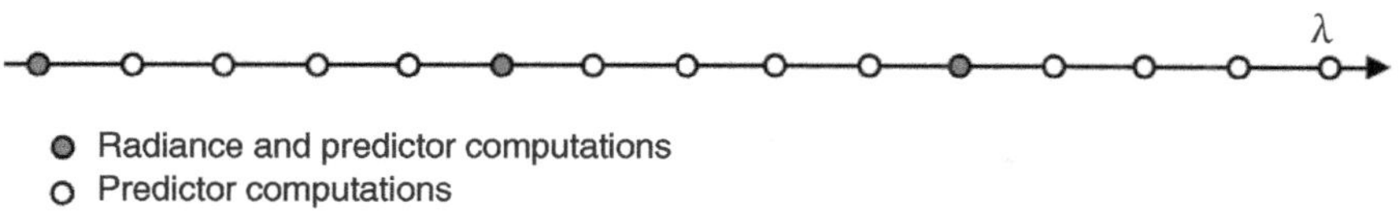

Figure 7.7 Illustration of the method. Here the sampling rate is $s = 5$. *Source*: Adapted from Efremenko et al. (2023) (licensed under CC BY 4.0).

Note that the third term $1/\lambda_i^4$ corresponding to Rayleigh scattering is added empirically to improve the results in the spectral range 400–600 nm. Assuming a linear model between the components of p_i and the radiance value $R_i = R(\lambda_i)$, we have:

$$R_i = aT_{\sigma i} + bT_{ki} + c\frac{1}{\lambda_i^4} + d. \tag{7.15}$$

Then we choose a spectral range containing K consequent spectral points from R_m array, for which the radiances are known, and make an overdetermined system of Eq. (7.15). The unknown coefficients a, b, c, and d are solved by a least square minimization. Finally, the values of radiances within a considered spectral range are computed using the found coefficients in Eq. (7.15). As shown in Figure 7.7, at the filled points, the full radiative transfer computations are performed, while the predictor computations based on optical parameters are evaluated both at white and filled spectral points. After the coefficients a, b, c, and d are found from the "filled" points, the radiances are restored at white points.
Note that by setting

$$R_i = ak_i + d, \tag{7.16}$$

we derive an analog of the k-distribution model, while setting

$$R_i = a\hat{R}_i + bT_{ki} + d, \tag{7.17}$$

with $\hat{R}_i$ being the radiance found by using the two-stream radiative transfer model, the low-stream regression model (del Águila and Efremenko, 2021) is obtained.

The described method has been applied to test scenarios. The restored spectra and the reference spectra have been convolved with the Gaussian spectral response function (SRF). The relative error is computed across the spectrum and averaged over geometries within each scenario type. The values of sampling rates are listed in Table 7.1. Figure 7.8 shows the examples of spectral computations for the clear and aerosol and cloud cases. For clear-sky and aerosol cases, SRF full width at half maximum (FWHM) is 0.04 nm and the sampling rate is $s = 2048$, while for the cloud scenario, the sampling rate is $s = 64$.

The errors of the method for the spectral range 400–1000 nm are summarized in Table 7.1. The error increases with the sampling rate. However, even for $s = 2048$, the mean error is below 1%. The maximum errors correspond to some absorption lines in the water vapor region around 950 nm. For the cloud case, the mean error is about 1% at the sampling rate $s = 128$, while the maximum error is larger than 15%. Also, note that the results for the cloud scenario are somewhat unstable since the maximum error for $s = 128$ is less than that for $s = 16$.

The results for the spectral range 1000–2500 nm are summarized in Table 7.2. Overall, the errors are lower than those for the 400–1000 nm range. It can be explained by the fact that

Table 7.1 The efficiency of the method when the SRF with FWHM = 50 nm is applied. Spectral range is 400–1000 nm.

Scenario	Sampling rate	Grid step	Mean/max error (%)
Clear sky	16	0.182	0.024/1.848
	128	1.459	0.055/3.037
	2048	23.347	0.64/2.958
	4096	46.694	1.445/14.288
Aerosol (AOT = 2)	16	0.182	0.058/2.202
	128	1.459	0.159/1.54
	2048	23.347	0.405/2.669
	4096	46.694	0.634/4.347
Cloud	16	0.182	0.318/21.989
	128	1.459	0.796/16.459
	2048	23.347	2.622/45.903
	4096	46.694	4.608/43.588

Table 7.2 The same as in Table 7.1, but for the spectral range 1000–2500 nm.

Scenario	Sampling rate	Grid step	Mean/max error (%)
Clear sky	16	0.48	0.05/49.031
	128	3.84	0.033/49.031
	2048	61.44	0.046/49.03
	4096	122.88	0.052/49.036
Aerosol (AOT = 2)	16	0.48	0.21/49.9
	128	3.84	0.192/50.0
	2048	61.44	0.48/51.88
	4096	122.88	0.62/50.3
Cloud	16	0.48	0.378/50.25
	128	3.84	0.97/49.924
	2048	61.44	1.37/51.526
	4096	122.88	4.48/51.20

the Rayleigh scattering cross-section is negligibly small in the infrared region and so, the impact of the scattering cross-section and multiple scattering is small. The maximum error (about 50%) is at 1400 nm where the absolute value of the radiance is close to zero. For clear and aerosol cases, the sampling rate $s = 2048$ keeps the mean error around 0.5%. For the cloud scenario, the sampling rate below $s = 128$ keeps the mean error below at least 1%.

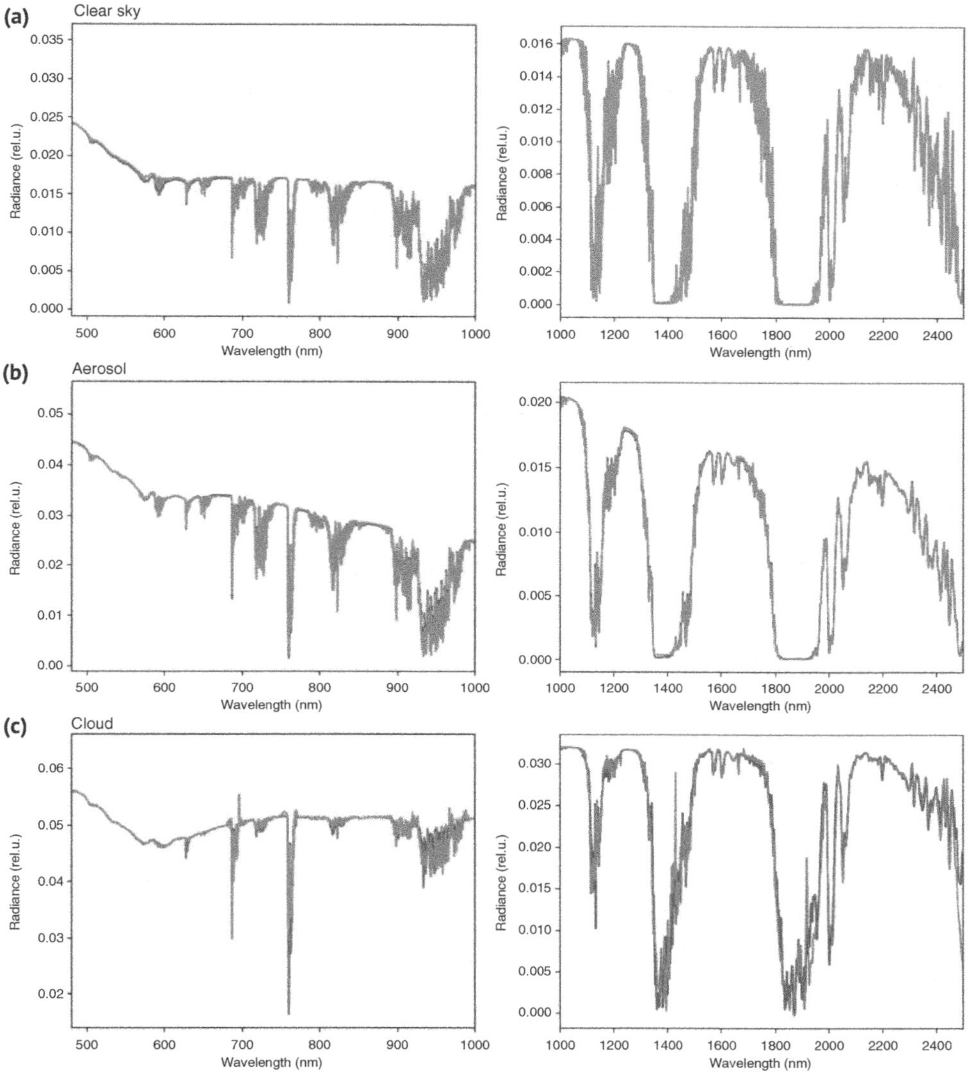

Figure 7.8 Radiance spectra for (a) clear-sky, (b) aerosol, and (c) cloud scenarios convolved with SRF with FWHM = 0.04 nm and the sampling rate $s = 2048$ ($s = 64$ for the cloud scenario). *Source*: Adapted from Efremenko et al. (2023) (licensed under CC BY 4.0).

A practical conclusion from our analysis is that as long as the aerosol loading is relatively low (below 2), the full radiative transfer simulations can be performed on a coarse wavelength grid with a step of about 20 nm in the 400–1000 nm range and 60 nm in 1000–2500 nm. In this case, using the described method, a high-resolution spectrum can be restored with an error lower than 0.5%. The simulations in the case of optically thick clouds require a finer discretization step. In fact, for this case, the principal component-based methods, considered further, look more promising.

7.7 Concept of Dimensionality Reduction

Dimensionality reduction means representing the initial data set with less number of parameters than it is initially represented. It can be considered as one of the lossy data compression paradigms. Dimensionality reduction is crucial for stable and high-performance processing of spectral measurements. It excludes redundant information from the initial dataset, reduces the number of independent parameters, and improves the efficiency of machine learning.

There is a distinction between linear and nonlinear techniques for dimensionality reduction. A more detailed review can be found in Fodor (2002), Gorban et al. (2008), and references therein. Linear and nonlinear techniques have been intercompared by van der Maaten et al. (2009). Results of these numerical experiments reveal that nonlinear techniques perform well on selected artificial tasks. However, they hardly outperform PCA on real-world tasks. Bearing in mind that no obvious superior method has emerged in the benchmarking studies (increasingly time-consuming and sophisticated dimensionality reduction techniques lead to more accurate results, and vice versa), our analysis will be limited with the classical PCA.

The theory of dimensionality reduction can be briefly summarized as follows (Efremenko et al., 2017a). Let $\mathbf{y} = (y(\lambda_1), y(\lambda_2), \ldots, y(\lambda_W))$, $\mathbf{y} \in \mathbb{R}^{1\times W}$, be a row vector of atmospheric radiances at W wavelengths $\{\lambda_w\}_{w=1,\ldots,W}$. A set of S spectra is assembled into a matrix $\mathbf{Y} \in \mathbb{R}^{S\times W}$ whose i-th row is y_i. Then, $\mathbf{y}_i$ can be represented in a new basis system as follows:

$$\mathbf{y}_i = \overline{\mathbf{y}} + \sum_{k=1}^{W} t_{ik}\mathbf{f}_k. \tag{7.18}$$

Here, $\overline{\mathbf{y}} = \sum_{i=1}^{S} \mathbf{y}_i/S$, $\overline{y} \in \mathbb{R}^{1\times W}$ is the sample mean of the spectra (the average spectrum), t_{ik} is the kth coordinate of the vector $\mathbf{y}_i$ in the new basis system, and $\mathbf{f}_k = (f_k(\lambda_1), f_k(\lambda_2), \ldots, f_k(\lambda_W)) \in \mathbb{R}^{1\times W}$ is the kth basis vector. Noting that high-dimensional real data are often situated on or near a lower-dimensional manifold, the spectrum $\mathbf{y}_i$ can be projected onto the K-dimensional subspace ($K < W$) as follows:

$$\mathbf{y}_i \approx \overline{\mathbf{y}} + \sum_{k=1}^{K} t_{ik}\mathbf{f}_k, \tag{7.19}$$

or in matrix form for the initial dataset:

$$\mathbf{Y} \approx \overline{\mathbf{Y}} + \mathbf{TF}, \tag{7.20}$$

where $\overline{\mathbf{Y}} = \{\overline{\mathbf{y}}, \ldots, \overline{\mathbf{y}}\} \in \mathbb{R}^{S\times W}$, $\mathbf{F} = \{\mathbf{f}_1, \ldots, \mathbf{f}_K\}$, $T \in \mathbb{R}^{S\times K}$ is the matrix whose entries are $\{t_{ik}\}_{i=1,\ldots,S}^{k=1,\ldots,K}$. Hereinafter, the superscript ^{T} stands for "transpose." The transformation (7.19) can be done using dimensionality reduction techniques, such as PCA (Pearson, 1901). In the latter, basic vectors $\mathbf{f}_k$ in Eq. (7.19) are referred to as "principal components" (PCs) or empirical orthogonal functions (EOFs), while the coordinates t_{ik} in the new coordinate system and the corresponding matrix $\mathbf{T}$ are called "principal component scores." Considering a spectral decomposition for the covariance matrix $\mathrm{cov}(\mathbf{Y}, \mathbf{Y}) \equiv C_{\mathrm{Y}} \in \mathbb{R}^{W\times W}$:

$$C_{\mathrm{Y}} = \mathbf{E}\Lambda\mathbf{E}^{\mathrm{T}}, \tag{7.21}$$

where $\mathbf{E} \in \mathbb{R}^{W \times W}$ is the eigenvector matrix and $\Lambda \in \mathbb{R}^{W \times W}$ is the diagonal matrix of eigenvalues, and the principal components are taken as K eigenvectors related to the K most significant eigenvalues. The principal component scores are then computed as follows:

$$\mathbf{T} = \left(\mathbf{Y} - \overline{\mathbf{Y}}\right)\mathbf{F}^{\mathrm{T}}. \tag{7.22}$$

7.8 Principal Component Analysis of Spectral Radiances

The most conceptually simple approach uses the training data set of spectra in order to establish a set of EOFs by using PCA and then to restore hyperspectral signal in W spectral points by using K monochromatic radiances. Naturally, we have $K<W$.

The theory of PCA briefly discussed in the previous section reveals a linear relationship between PC scores and monochromatic radiances:

$$y(\lambda) = \bar{y}(\lambda) + t_1 f_1(\lambda) + \ldots + t_K f_K(\lambda). \tag{7.23}$$

Hence, for a given set of K EOFs and K spectral points, it is possible to obtain a closed linear system of K equations:

$$\begin{cases} y(\lambda_1) = \bar{y}(\lambda) + t_1 f_1(\lambda_1) + \ldots + t_K f_K(\lambda_1), \\ y(\lambda_2) = \bar{y}(\lambda) + t_1 f_1(\lambda_2) + \ldots + t_K f_K(\lambda_2), \\ \ldots \\ y(\lambda_K) = \bar{y}(\lambda) + t_1 f_1(\lambda_K) + \ldots + t_K f_K(\lambda_K). \end{cases} \tag{7.24}$$

The key point here is that the radiance values in K spectral points are represented by the same EOFs. Then, by solving Eq. (7.24), we obtain PC scores $t_1, \ldots t_K$, and, by using Eq. (7.23), the full spectrum in W spectral points can be readily restored.

This approach requires a set of precomputed EOFs derived from a training data set of simulated or measured spectra. Figure 7.9 shows an example of the PCA applied to the dataset of spectra computed in the Hartley–Huggins band used for ozone retrieval. The data set consists of 10^5 spectra. The following parameters are varied for the generation of reflectance spectra: the solar zenith angle, the viewing zenith angle, the relative azimuthal angle, the surface albedo, the ozone total column, the surface height, and the temperature. The right plot in Figure 7.10 shows that almost 99.9% of the variance in the data can be explained just with five principal components.

To obtain best efficiency, the parameters of this scheme, such as K and a set of chosen wavelengths, have to be tuned empirically. The number K depends on the desired level of variance to be captured by the principal components. Several semi-empirical rules have been proposed for the optimal number of principal components (e.g. the broken stick model[1] (MacArthur, 1957)). However, there is no universal rule for the selection of K; the choice

1 The broken stick model in PCA helps determine the number of principal components to retain by comparing their explained variance to a theoretical distribution of variance under random conditions. The idea is that if the eigenvalues from the PCA are larger than those expected under the broken stick distribution, the corresponding principal components explain more variance than would be expected by chance.

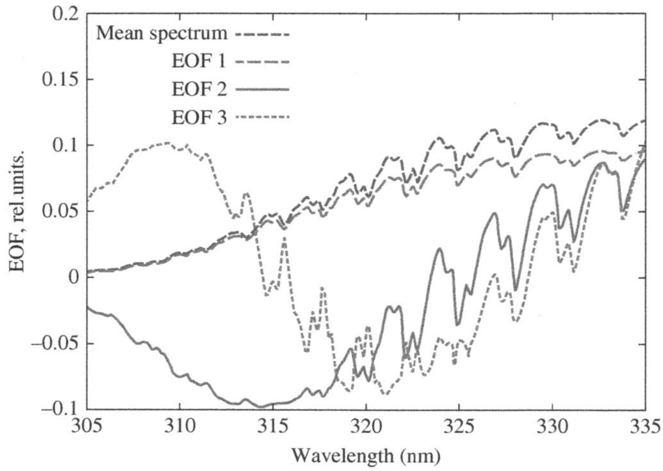

Figure 7.9 Mean spectrum and first three EOFs computed in the Huggins band. *Source*: Adapted from del Águila et al. (2019a) (licensed under CC BY 4.0).

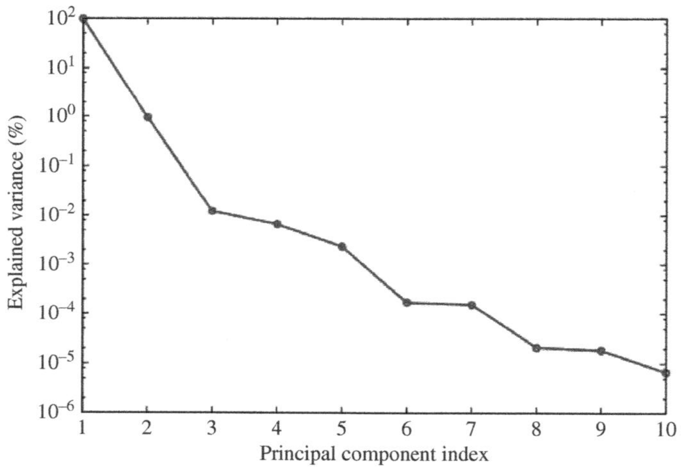

Figure 7.10 Explained variance in percentage as a function of the principal component index. *Source*: Adapted from del Águila et al. (2019a) (licensed under CC BY 4.0).

is application-specific. Regarding the choice of wavelengths, a method is proposed by Liu et al. (2006) for selecting the location of monochromatic wavelengths by using a correlation function. In particular, this method involves the following steps:

1. The correlation coefficients are computed for the radiance values and then converted to vector angles by an arccosine function.
2. The spectral data is rearranged according to the magnitudes of the correlation coefficients.
3. The monochromatic radiances are selected by choosing predictors with equal distances in the values of the correlation coefficients.

The schematic representation of the principal component-based hyperspectral radiative transfer model is shown in Figure 7.11. For an input data set containing the optical parameters of the atmosphere for a set of wavelengths, the monochromatic radiative transfer solver is executed, producing a spectral data set, which is divided into training data set and validation data set. By applying PCA and the correlation analysis to the training set, the system of EOFs is computed, and a subset of spectral points is chosen (spectral sampling), respectively. These two outputs are stored and used for computing PC scores for the validation data set. The spectra in the full wavelength range are restored using Eq. (7.23), and the error of this reconstruction can be estimated. If the error is larger than required, the number of generated spectra and the number of principal components are increased. The main outputs of the training phase are EOFs and spectral sampling (as shown in Figure 7.11), which allow to process new data in the online phase (as shown in Figure 7.12).

In Liu et al. (2006), it was noted that the slit function convolution operator and the PCA are linear. Therefore, the PC scores of the convolved spectra are linear functions of monochromatic radiances. Then at the training stage, the corresponding weighting factors of linear dependency are stored together with the system of EOFs and spectral sampling.

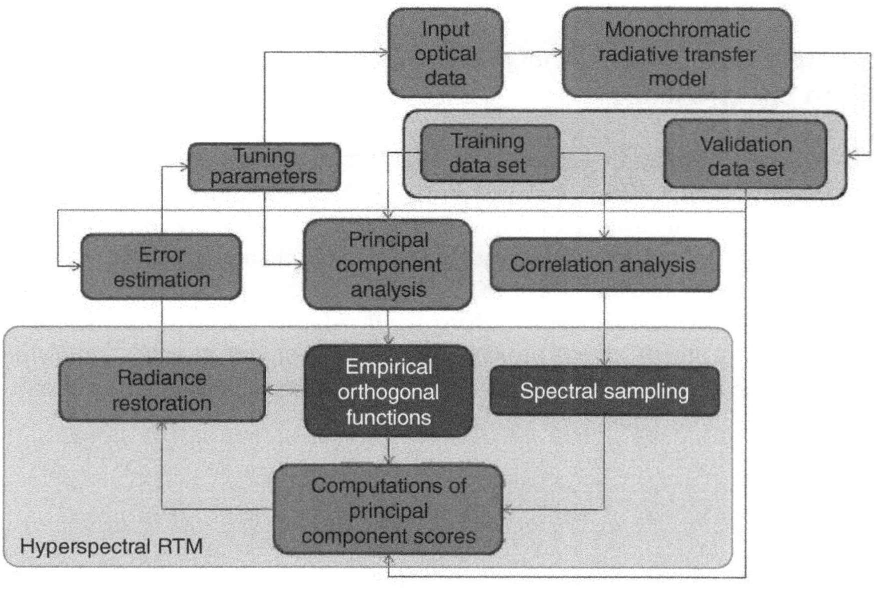

Figure 7.11 Schematic representation of the PCA-based radiative transfer model with precomputed EOFs – offline phase. *Source*: Adapted from del Águila et al. (2019b).

Figure 7.12 Same as Figure 7.11, but showing the online phase. *Source*: Adapted from del Águila et al. (2019b).

Authors claim that for an infrared spectrum ranging from 650 to 3000 cm^{-1}, this representation for convolved spectra reduces the number of monochromatic computations from a few thousands to a few hundreds.

Although the presented approach proved to be efficient and implemented in several packages (e.g. PCTRM (Liu et al., 2006), RTTOV (Matricardi, 2010), and others (Hurley et al., 2012; Hollstein and Lindstrot, 2014)), the main drawback is that it still requires time-consuming training data set computations.

Following this methodology, Liu et al. (2006) used PCA to compress channel transmittances or radiances in the infrared spectral range. They found that 100–250 PCs were sufficient to reproduce channel radiances to 0.01 K accuracy for hyperspectral sensors with thousands of channels. The EOFs capture the spectral variations of the radiances, while the PC scores capture the temperature and composition dependence. An ensemble of synthetic spectra was used to generate the EOFs and PC scores. Since the EOFs and the instrument line shape (ILS) do not depend on the atmospheric state, the PC scores for other scenarios can be obtained as projections of the channel radiance on the EOFs. The channel radiance is a linear combination of monochromatic radiances within the frequency range of that channel, with the weights being the normalized ILS at the frequency grid point. Liu et al. (2006) use a correlation function to select the location of the monochromatic grid points.

7.9 Principal Component Analysis for Differential Optical Absorption Spectroscopy

In this section, we consider the application of PCA in the framework of the differential optical absorption spectroscopy (DOAS) (Platt and Stutz, 2008). DOAS is widely used to retrieve trace gas content. The main advantages of this technique are its simplicity and robustness, as it can filter out the influence of factors that are not taken into account properly in the radiative transfer model. The main idea of DOAS lies in the fact that absorption caused by gases leads to strong spectral signatures in the spectral radiances, while the influence of multiple scattering and other factors is usually smooth in the wavelength space. Following Platt (1994), the radiance reflected by the atmosphere consisting of N_g gas species is represented using the weak absorption Beer–Lambert law as

$$\ln y(\lambda) = -\sum_{g=1}^{N_g} S_g \sigma_g(\lambda) - P(\lambda) - \mathrm{RRS}(\lambda), \tag{7.25}$$

where S_g is the number density of gas g along the optical path (also referred to as the slant column density), $P(\lambda)$ is the polynomial term that represents the impact of Rayleigh and aerosol/cloud scattering as well as the surface reflectance, and RRS is a term representing the rotational Raman scattering (Kattawar et al., 1981). In the conventional DOAS, S_g is retrieved through least squares fitting that minimizes the residual between the measured (left part of Eq. (7.25)) and simulated (right part of Eq. (7.25)) radiance spectra. Then S_g in converted into the vertical column density (Ω_g) though the air mass factor. The latter is computed at a single wavelength assuming a prescribed vertical profile of gas g (Perliski and Solomon, 1993; Slusser et al., 1996).

In Li et al. (2013), a modification of the DOAS approach was proposed for retrieving the SO_2 total column. The PCA is applied for the measured spectra in regions with no significant SO_2, e.g. the equatorial Pacific:

$$\ln y(\lambda) = \overline{\ln y}(\lambda) + \sum_{i=1}^{K} t_k f_k(\lambda). \tag{7.26}$$

In this way, the EOFs capture the variability of the data caused by physical processes (i.e. Rayleigh and Raman scattering and ozone absorption). In addition, the features of the instrument (e.g. the instrumental degradation, the slit function, and measurement artifacts) are implicitly accounted for by the EOFs. That is the training phase. Then, for polluted regions with SO_2, Eq. (7.26) will produce a residual associated with SO_2 content. Thus,

$$\ln y(\lambda) = \overline{\ln y}(\lambda) + \sum_{i=1}^{K} t_k f_k(\lambda) + \Omega_{SO_2} \frac{\partial \ln y(\lambda)}{\partial \Omega_{SO_2}}. \tag{7.27}$$

The derivative in the last term can be estimated either by finite differences or by using linearized radiative transfer models considered by Spurr et al. (2013) and Efremenko et al. (2014). Then, from Eq. (7.27), the vertical column density (Ω_{SO_2}) can be readily retrieved.

This method has been applied to the Ozone Monitoring Instrument (Levelt et al., 2006) data in the spectral interval 310–340 nm. As the high-order principal components represent noise rather than a useful signal, the truncation over the principal components also acts as a filter. To reconstruct the spectral radiances, at least 20–30 principal components were required, while in the presence of relatively strong SO_2 signals, that number could be reduced to 8. Authors claim that the noise in the data was decreased by a factor of 2, thereby providing greater sensitivity to anthropogenic sources of SO_2.

So far, there are no reports of applying a similar approach to other trace gases. One reason for that is the difficulty of obtaining the system of EOFs for regions without a certain trace gas. The second reason is that, strictly speaking, Eq. (7.27) is approximate. For SO_2, it works correctly, and the residual is associated with the SO_2 signal. For other trace gases, Eq. (7.27) might not be valid and more elaborated approach is required.

7.10 Principal Component Analysis of Optical Parameters

Natraj et al. (2005) used principal component analysis (PCA) of inherent optical properties (such as optical depth and single-scattering albedo) to enhance the computational efficiency of LBL calculations. PCA employs orthogonal basis functions to reduce the dimensionality of data sets. Natraj et al. (2010) extended this approach to include polarization effects. The PCA-based approach gains its speed through reductions in

1. the number of wavelengths at which RT calculations are performed, and
2. the time taken for each such RT calculation.

Natraj et al. (2005) used a three-step process for wavelength optimization: binning, eigenvalue problem solution, and radiance back-mapping. For a window with thousands of spectral points, there is invariably a large dynamic range of optical depths; the window is therefore partitioned into a number of spectral bins. Each bin is characterized by grouping certain optical properties (such as cumulative atmospheric trace gas optical depths) that are similar within the bin. The trace gas absorption optical depth is used rather than the total extinction optical depth because the former has much more significant spectral variation than that due to scattering by aerosols, clouds, and air molecules. Furthermore, the aerosol loading can vary substantially between scenes, and it is desirable not to change the binning parameters from scene to scene. These bins could also be further divided by splitting along the median scattering optical depth.

The PCA application for RT performance enhancement exploits redundancy in a binned data set of optical properties by expressing the variance of this data set in terms of the first few EOFs. Typically, four EOFs capture more than 99.99% of the variance in the optical properties. These EOFs define a much smaller set of "PCA-projected" optical states. Multiple-scattering radiative transfer calculations are performed only for this reduced set of optical states. In this manner, the following quantities can be computed:

$$I_d^0 = \ln(I^0/I_2^0), \tag{7.28}$$

$$I_{d,k}^{\pm} = \ln(I_k^{\pm}/I_{2,k}^{\pm}), \tag{7.29}$$

where k is the EOF index, subscript 2 denotes two-stream radiances, superscripts 0, +, and − refer to calculations for bin mean value, and positive and negative EOF perturbations from the mean, respectively. I^0 and $I_k^{\pm}$ indicate high-accuracy, multiple-scattering calculations for the bin mean and EOF-perturbed states, respectively.

Radiance restoration is based on the assumption that, for all wavelengths in the given bin, the difference between the logarithm of the accurate (multiple-scattering) radiance field and the approximate (two-stream) field can be derived from the PCA-projected RT results and a second-order central-difference expansion in terms of the principal component projections.

$$I_l = I_{SS} + I_2 \exp\left[I_d^0 + \sum_k \delta I_k P_{kl} + \frac{1}{2}\sum_k \delta^2 I_k P_{kl}^2\right], \tag{7.30}$$

where l is the wavelength index; SS denotes a single-scattering calculation; δI_k and $\delta^2 I_k$ are, respectively, the first- and second-order central intensity differences for EOF k; and P_{kl} refers to the projection of the original data set onto EOF k for wavelength index l.

PCA is similar to the double-k, linear-k, and LSI approaches discussed in Section 7.5, in that an exact single-scattering calculation is performed at every wavelength, with PCA used only to evaluate the multiple-scattering radiance field. It is also worth noting that the second-order Taylor expansion employed for the radiances likely implies first-order accuracy of the Jacobians.

Note that PCA is done on logarithms of optical properties, and the radiation-field restoration likewise uses log-space. The use of logarithms for the intensity calculation offers two advantages: negative intensities are automatically avoided, and the fact that extinction is exponential in the absence of scattering is implicitly taken into account. Employing two-stream calculations reduces the number of terms needed in the PCA expansion for a given accuracy requirement. This is related to the fact that PCA is a linear transformation. Considering scattering as a perturbation to gaseous absorption and using the fact that the

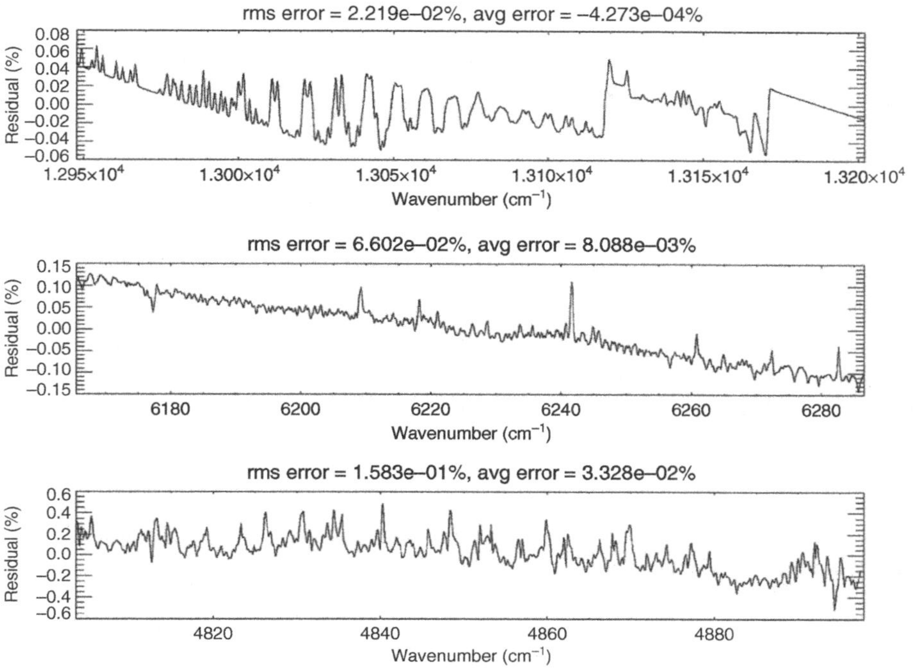

Figure 7.13 Error in the radiance for PCA calculations compared to LBL calculations in the (top) O_2 A-band; (middle) 1.61 μm CO_2 band; (bottom) 2.06 μm CO_2 band. The error is defined as (PCA−LBL)/LBL×100. *Source*: Used with permission from Natraj et al. (2010). © Elsevier.

two-stream approximation is exact in the absence of scattering, it is evident that the ratio of high-accuracy multiple stream and two-stream intensities is almost linearly dependent on optical properties, making the PCA procedure very efficient.

Figure 7.13 shows the percentage difference in the radiance between PCA and LBL computations for an example scenario with aerosols, ice, and water cloud scattering. The top, middle, and bottom panels denote the O_2A-band, the 1.61 μm CO_2 band, and the 2.06 μm CO_2 band spectral regions, respectively. Two interesting features are evident. First, the results demonstrate a clear slope. This is a consequence of averaging (wavelength-dependent) aerosol/cloud scattering properties within each bin. The slope can be removed through the addition of a simple parameter related to column scattering optical depth in the PCA process. Second, errors in the 2.06 μm CO_2 band are larger than the corresponding values in the other spectral regions. This is because the 2.06 μm CO_2 band has two significant absorbers (H_2O and CO_2), as opposed to the single dominant absorber in the other bands. This effect can be mitigated by adding a few additional bins for lines with large H_2O absorption, at the cost of a small increase in computation time.

del Águila et al. (2019a) introduced a unified methodology whereby PCA was applied in both the optical parameter and spectral radiance spaces. It was found that for the Hartley–Huggins band, the combined application of optical property and radiance-based PCA yields an accuracy better than 0.05% with a speedup of about 20 compared to LBL calculations.

7.11 Neural Networks

Application of neural networks to atmospheric remote sensing has been a topic of research for more than 25 years. Key and Schweiger (1998) emulated a medium spectral resolution radiative transfer model using neural network techniques, with application to the computation of surface fluxes. Schwander et al. (2001) developed a neural network method for increasing the computational efficiency of high spectral resolution radiative transfer models. The basic idea is for an accurate radiative transfer model to calculate radiances at a few (M) selected wavelengths, and for the neural network algorithm to emulate a complete high-resolution spectrum comprising N ($\gg M$) wavelengths.

The schematic structure of a neural network is shown in Figure 7.14. The input layer has $M + a$ neurons, where a is the number of auxiliary quantities (such as the solar zenith angle and the total ozone amount). The input vector $\mathbf{x}$ includes the transmittances at the M wavelengths, along with the auxiliary quantities. The output vector $\mathbf{o}$ consists of the transmittances at the N wavelengths. Typically, one hidden layer with J neurons is used to connect the input and output layers. Neurons in one layer are not connected to each other, but are connected to all neurons in adjacent layers, making the neural network "fully interconnected." Weights (w_{ij} and w_{jk}) are used to propagate the signal from a neuron in one layer to that in another.

The neural network is trained by propagating the input data successively through it. The resulting output vector is then compared with the target vector. The weights of the neural network are iteratively refined based on the error between the output and target vectors. One epoch is completed when the entire training set is passed through the neural network. At this point, the procedure starts over, and the steps are repeated. Training stops when the mean difference between the target and output vectors does not decrease anymore.

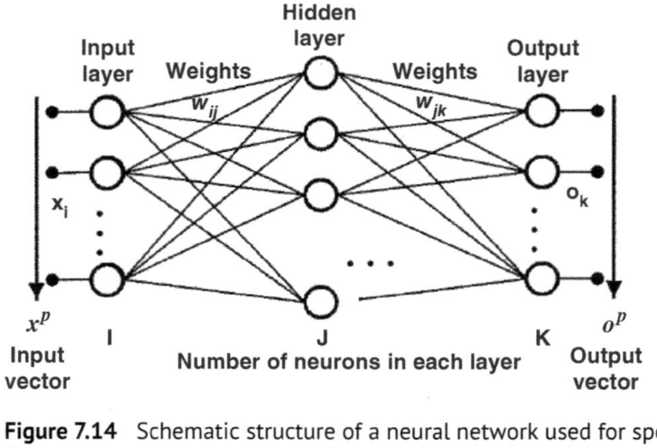

Figure 7.14 Schematic structure of a neural network used for spectral radiative transfer modeling. To train the neural network, the input data are propagated successively through the network. The resulting output vector is compared with the target vector. From the resulting error between the output and the target vector, the weights of the neural network are refined. In this manner, the entire training set is passed through the neural network to complete one epoch. Training continues until the mean difference between the target and output data stops decreasing. Finally, to test the performance of the neural network, a completely different set of spectra is used as a test data set and passed through the network. During this phase, the weights are not adapted. *Source*: Reprinted with permission from Schwander et al. (2001). © Optica Publishing Group.

An independent data set is passed through the network to test its performance and generality; the weights are not adapted during this test.

The algorithm was evaluated for an application involving the ultraviolet and visible spectral regions. The input data comprises transmittances at seven wavelengths (290, 300, 310, 342, 380, 470, 610 nm), together with the solar zenith angle and the total column ozone. The corresponding output data are high-resolution transmittances at 153 wavelengths in the 280–700 nm spectral range. A total of 20 394 spectra were provided for training, with another 20 934 spectra forming the test set. The transmittances were transformed into UV indices (that act as a measure of the intensity of ultraviolet radiation). Comparisons were made between results from the accurate and neural network-based radiative transfer models.

The ratio of UV index computed with and without the use of the neural network model is shown in the top panel of Figure 7.15 for the training data set and in the bottom panel for the test data set. The ratio deviates by less than 5% from unity in both cases, with most of the errors within 1%. The computation time is reduced by a factor of more than 20.

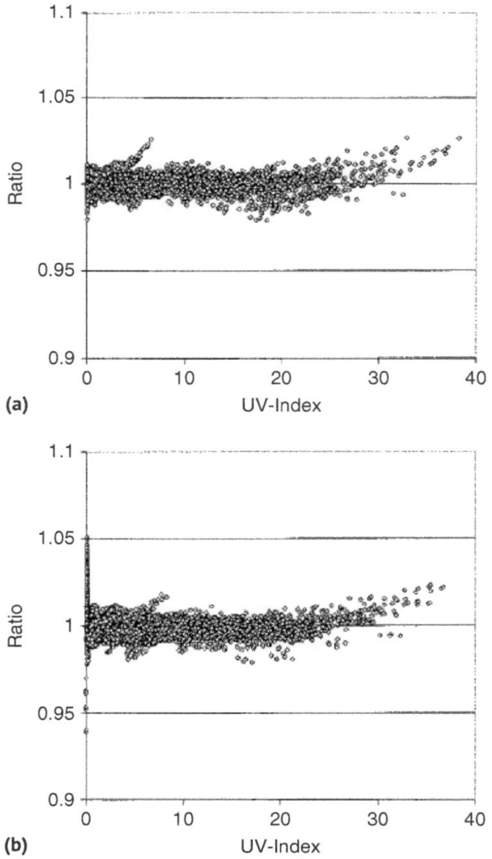

Figure 7.15 Ratio of UV index calculated with and without use of the neural network algorithm for (a) training data set and (b) test data set. *Source*: Reprinted with permission from Schwander et al. (2001). © Optica Publishing Group.

Recently, Takenaka et al. (2011) applied a neural network algorithm to data from the Multi-functional Transport Satellite-1 Replacement (MTSAT-1R) geostationary satellite and validated their results against *in situ* observations at four SKYNET sites. The method was also applied to observations from the Advanced Earth Observing Satellite-II/Global Imager (ADEOS-II/GLI).

In Loyola et al. (2018), it was reported that the neural network approach reproduced spectra with a mean absolute error below 1% over the O_2 A-band. In Bue et al. (2019), neural networks were used to reproduce data acquired by the PRISM imaging spectrometer (Mouroulis et al., 2014) designed to observe coastal ocean environments in the 350–1050 nm spectral range at approximately 3 nm spectral sampling. The neural network took 5 ms per spectrum, while computations based on monochromatic radiative transfer models took 10 minutes per spectrum, resulting in a speedup of five orders of magnitude.

Despite its high computational efficiency, the neural network-based approach has certain drawbacks, which can be summarized as follows:

- Extensive computations are required for the training data sets.
- Training is restricted to a chosen atmospheric model with a predefined number of atmospheric layers; in broadband computations, the slit function of the instrument is also fixed.
- Neural networks might not capture and reproduce implicit properties of the radiance field, such as the symmetry relation with respect to incident and viewing polar angles.
- In practice, the neural network-based approach is not fail-safe, i.e. errors can be significant in some edge cases that have not been captured during training.

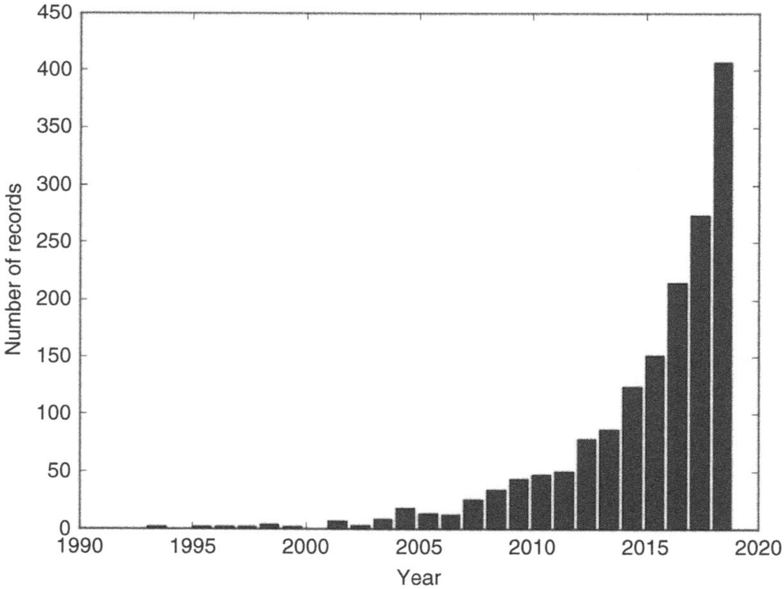

Figure 7.16 Number of records in the Web of Knowledge database corresponding to the topics of "machine learning" and "remote sensing."

In this regard, Efremenko (2021) explored a novel concept of neural network-based radiative transfer models. Instead of replacing the entire model, the neural network substitutes only a specific component, namely, the eigenvalue solver. This approach reduces computational time while preserving the generality of the model. The training excludes explicit dependencies on observation geometry and the optical thickness of the medium. While the speedup achieved with this method is modest (approximately a factor of 3), especially when compared to the thousand-fold improvements reported in recent studies, the neural network remains flexible and easy to train. It is also adaptable for any number of atmospheric layers.

Another neural network concept that may be used in radiative transfer is Physics-Informed Neural Networks (PINNs). This is a novel machine-learning approach designed to approximate solutions for both forward and inverse problems. Unlike traditional neural networks, PINNs incorporate the governing physical equations directly into the training process. This is achieved by minimizing the residuals of the underlying radiative transfer equations during training (Lagaris et al., 1998). The key advantage of PINNs lies in their ability to leverage physical laws, which enables them to provide fast, robust, and accurate solutions irrespective of the equation's dimensionality. PINN was utilized for solving forward and inverse problems of radiative transfer in Mishra and Molinaro (2021) and Zucker et al. (2025).

Several new retrieval algorithms are being developed that are suited to processing large amounts of remote sensing data in near-real time, as evidenced by the growing number of publications in this field (see Figure 7.16).

A

Legendre Polynomials

The associated Legendre polynomials $P_l^m(\mu)$ can be computed recurrently. The base of the recurrence is given by two formulas for $m = l$ and $l = m + 1$, respectively,

$$P_m^m(\mu) = (-1)^m (2m-1)!!(1-\mu^2)^{m/2} \tag{A.1}$$

and

$$P_{m+1}^m(\mu) = \mu(2m+1)P_m^m(\mu), \tag{A.2}$$

where !! denotes the double factorial.[1] Polynomials $P_l^m(\mu)$ for $l \geq m + 2$ are computed by using a three-term recurrence relation with respect to the degree:

$$(2l+1)\mu P_l^m(\mu) = (l+m)P_{l-1}^m(\mu) + (l-m+1)P_{l+1}^m(\mu). \tag{A.3}$$

For validation of the recurrent formula implementation, it is convenient to use the analytical forms of the first few associated Legendre polynomials, which are listed in Table A.1.

The orthogonal property for $P_l^m(\mu)$ reads as follows:

$$\int_{-1}^{1} P_l^m(\mu)P_{l'}^m(\mu)d\mu = \frac{2}{2l+1}\frac{(l+m)!}{(l-m)!}\delta_{ll'} \tag{A.4}$$

Table A.1 The first few associated Legendre polynomials.

	$l = 0$	$l = 1$	$l = 2$	$l = 3$
$m = 0$	1	μ	$\frac{1}{2}(3\mu^2 - 1)$	$\frac{1}{2}\mu(5\mu^2 - 3)$
$m = 1$		$-(1-\mu^2)^{1/2}$	$-3\mu(1-\mu^2)^{1/2}$	$\frac{3}{2}(1-5\mu^2)(1-\mu^2)^{1/2}$
$m = 2$			$3(1-\mu^2)$	$15\mu(1-\mu^2)$
$m = 3$				$-15(1-\mu^2)^{3/2}$

1 The double factorial, denoted by $n!!$, is a mathematical operation that extends the concept of the regular factorial. The double factorial applies to both odd and even integers and is defined as the product of all integers of the same parity (odd or even) from n down to 1 (or 2 for even numbers).

Analytical Methods in Radiative Transfer, First Edition. Alexander Kokhanovsky, Vijay Natraj and Dmitry Efremenko.

and the maximum value of $P_l^m(\mu)$ rapidly increases with m. To avoid overflow during recurrent process, it is recommended to compute normalized Legendre polynomials $\bar{P}_l^m(\mu)$ defined as

$$\bar{P}_l^m(\mu) = \sqrt{(2l+1)\frac{(l-m)!}{(l+m)!}}P_l^m(\mu) \tag{A.5}$$

instead of $P_l^m(\mu)$. For $\bar{P}_l^m(\mu)$, the following normalization condition holds:

$$\int_{-1}^{1} \bar{P}_l^m(\mu)\bar{P}_{l'}^m(\mu)d\mu = 2\delta_{ll'}. \tag{A.6}$$

B

Computations of Local Optical Parameters

Mie theory (Mie, 1908), formulated by Gustav Mie in 1908, provides a rigorous solution to the scattering and absorption of electromagnetic waves by spherical particles. It is based on Maxwell's equations within spherical coordinates Mie theory is applied under specific boundary conditions that ensure physically realistic and mathematically stable solutions (Bohren and Huffman, 1998; van de Hulst, 1957).

The first boundary condition requires that the electric (**E**) and magnetic (**H**) fields are continuous across the surface of the sphere. This condition is expressed as:

$$\mathbf{E}_{\text{inside}}|_{\text{surface}} = \mathbf{E}_{\text{outside}}|_{\text{surface}},$$

$$\mathbf{H}_{\text{inside}}|_{\text{surface}} = \mathbf{H}_{\text{outside}}|_{\text{surface}}.$$

This ensures that there are no discontinuities in the field components at the interface between the sphere and its surrounding medium.

The second boundary condition stipulates that, at infinity, the electric and magnetic fields should resemble spherical waves, decaying as $\frac{1}{r}$, where r is the distance from the center of the sphere. This decay is crucial because:

$$\lim_{r\to\infty} (\mathbf{E}, \mathbf{H}) \propto \frac{1}{r}. \tag{B.1}$$

If this condition were not met, the Poynting vector, which represents the flow of electromagnetic energy, would decay more slowly than $\frac{1}{r^2}$, leading to divergence in the energy integral across the sphere.

Considering that the scattered wave is proportional to the incident wave and behaves as a spherical wave in the far-field region, we can formulate the relationship as follows:

$$\begin{pmatrix} E_{\text{l}}^{\text{s}} \\ E_{\text{r}}^{\text{s}} \end{pmatrix} = \frac{e^{-ikr+ikz}}{ikr} \begin{pmatrix} S_2 & S_3 \\ S_4 & S_1 \end{pmatrix} \begin{pmatrix} E_{\text{l}}^{\text{i}} \\ E_{\text{r}}^{\text{i}} \end{pmatrix}, \tag{B.2}$$

where the superscripts s and i denote the scattered and incident waves, respectively. The subscripts l and r refer to the electric field components perpendicular and parallel to the plane of scattering. The matrix elements S_1, S_2, S_3, and S_4 are the angular-dependent

Analytical Methods in Radiative Transfer, First Edition. Alexander Kokhanovsky, Vijay Natraj and Dmitry Efremenko.

amplitude functions. For spherical particles, it is noted that $S_3 = 0$ and $S_4 = 0$ due to symmetry considerations, simplifying the scattering matrix to:

$$\begin{pmatrix} E_l^s \\ E_r^s \end{pmatrix} = \frac{e^{-ikr+ikz}}{ikr} \begin{pmatrix} S_2 & 0 \\ 0 & S_1 \end{pmatrix} \begin{pmatrix} E_l^i \\ E_r^i \end{pmatrix}. \tag{B.3}$$

The derivation of Mie solutions for the scattering of electromagnetic waves by spherical particles involves a crucial step: the expansion of the incident field, the scattered field, and the field inside the particle into a series of vector spherical wave functions (VSWFs). VSWFs form a complete basis for solutions to Maxwell's equations, meaning any function expressed as a linear combination of VSWFs automatically satisfies these equations. Specifically:

- The **incident and internal fields** are expressed in terms of *regular* VSWFs, ensuring the fields remain bounded and finite within the particle.
- The **scattered field** is expressed using *radiating* VSWFs, which are designed to meet the boundary conditions at infinity, depicting outgoing waves.

The Mie solution can be represented as two infinite series:

$$S_1(\Theta) = \sum_{n=1}^{\infty} \frac{2n+1}{n(n+1)} \{a_n \pi_n(\cos\Theta) + b_n \tau_n(\cos\Theta)\}, \tag{B.4}$$

$$S_2(\Theta) = \sum_{n=1}^{\infty} \frac{2n+1}{n(n+1)} \{b_n \pi_n(\cos\Theta) + a_n \tau_n(\cos\Theta)\}, \tag{B.5}$$

where π_n and τ_n are angular functions related to the associated Legendre functions:

$$\pi_n(\cos\Theta) = \frac{P_n^1(\cos\Theta)}{\sin\Theta}, \tag{B.6}$$

$$\tau_n(\cos\Theta) = \frac{dP_n^1(\cos\Theta)}{d\Theta}. \tag{B.7}$$

Equations (B.4) and (B.5) can be regarded as a multipole expansion of the scattered light, where coefficients a_n and b_n correspond to electric and magnetic multipole radiations, respectively. Specifically, a_n coefficients are related to electric dipole, quadrupole, and octupole radiation, while b_n coefficients are associated with their magnetic counterparts.

By considering the boundary condition at the particle surface and applying orthogonality properties of VSWFs, the unknown expansion coefficients a_n and b_n can be found (see full derivation in Bohren and Huffman (1998); van de Hulst (1957)). The Mie solution depends on two parameters. The first is the refractive index m of the particle with respect to the surrounding medium. The second parameter is the so-called size parameter defined as

$$x = \frac{2\pi r}{\lambda}, \tag{B.8}$$

where r is the particle radius, and λ is the wavelength. The final expressions for a_n and b_n read as follows:

$$a_n = \frac{\psi_n'(y)\psi_n(x) - m\psi_n(y)\psi_n'(x)}{\psi_n'(y)\zeta_n(x) - m\psi_n(y)\zeta_n'(x)}, \tag{B.9}$$

$$b_n = \frac{m\psi'_n(y)\psi_n(x) - \psi_n(y)\psi'_n(x)}{m\psi'_n(y)\zeta_n(x) - \psi_n(y)\zeta'_n(x)}, \tag{B.10}$$

where $y = mx$, primes denote the derivatives and

$$\psi_n(z) = zj_n(z), \tag{B.11}$$

$$\zeta(z) = zh_n^{(2)}(z) \tag{B.12}$$

with j and h^2 being spherical first and third kind Bessel functions (the latter is referred to as Hankel functions).

The extinction and scattering coefficients and also the phase function $p(\theta)$ for turbid media with spherical particles characterized by the size distribution $f(a)$ can be presented as

$$k_{\text{ext}} = N\int_0^\infty C_{\text{ext}}(a)f(a)da, \tag{B.13}$$

$$k_{\text{sca}} = N\int_0^\infty C_{\text{sca}}(a)f(a)da, \tag{B.14}$$

$$p(\theta) = \frac{2\pi N}{\kappa^2 k_{\text{sca}}}\int_0^\infty (|S_1(\theta)|^2 + |S_2(\theta)|^2)f(a)da, \tag{B.15}$$

where N is the number of particles in unit volume of a turbid medium. Essentially, C_{sca} and C_{abs} represent the electromagnetic powers removed from the incident wave as a result of scattering and absorption of the incident radiation, respectively, while C_{ext} gives the total electromagnetic power removed from the incident wave by the combined effect of scattering and absorption. The optical cross-sections have the dimension of area and depend on the direction and polarization state of the incident wave as well as on the size, optical properties, and orientation of the particle. The efficiencies (or efficiency factors) for extinction, scattering, and absorption are defined as:

$$Q_{\text{ext}} = \frac{C_{\text{ext}}}{G}, \quad Q_{\text{scat}} = \frac{C_{\text{scat}}}{G}, \quad Q_{\text{abs}} = \frac{C_{\text{abs}}}{G}, \tag{B.16}$$

where G is the particle cross-sectional area projected onto a plane perpendicular to the incident beam. In view of the definition of the normalized differential scattering cross-section, we set $G = \pi r^2$. From the point of view of geometrical optics, we expect that the extinction efficiency of all particles would be identically equal to unity.

The expressions for the extinction cross-section Q_{ext}, scattering cross-section Q_{sca}, and average cosine of scattering angle g for monodispersed spheres expressed via amplitude coefficients (B.9) and (B.10) are given by:

$$Q_{\text{ext}} = \frac{2}{x^2}\sum_{n=1}^{\infty}(2n+1)\mathbf{Re}(a_n + b_n), \tag{B.17}$$

$$Q_{\text{sca}} = \frac{2}{x^2}\sum_{n=1}^{\infty}(2n+1)(|a_n|^2 + |b_n|^2), \tag{B.18}$$

$$gQ_{\text{sca}} = \frac{4}{x^2} \sum_{n=1}^{\infty} \left[\text{Re}(a_n a_{n+1}^* + b_n b_{n+1}^*) \frac{2n+1}{n(n+1)\text{Re}(a_n b_n^*)} \right]. \tag{B.19}$$

One can see that the calculations of local optical characteristics of turbid media are quite complex and require the summation of infinite series and the application of the respective integration procedures. The calculations for the case of nonspherical particles are even more involved and require large amounts of computational time. Therefore, the databases of the precalculated or measured parameters are used in practical work. Here are some references to databases: Baum et al. (2014), Gao et al. (2022), Eriksson et al. (2018), Meng et al. (2010), Munoz et al. (2012), Hess et al. (1998), Yang et al. (2005), and Saito et al. (2021).

Alternatively, the simplified models for the case of special type of particles can be used, e.g. optically soft particles, particles that are much smaller or larger as compared to the wavelength of the incident light, etc. (van de Hulst, 1980). In particular, the following approximations for the extinction, scattering coefficients, and the parameter asymmetry of spherical particles, which are much larger as compared to the wavelength of the incident light ($x \to \infty, 2x|m-1| \to \infty$), can be used (Kokhanovsky, 2001; Kokhanovsky et al., 2024):

$$C_{\text{ext}} = 2\pi r^2, \tag{B.20a}$$

$$C_{\text{sca}} = \pi r^2 \left[1 + \rho + (1-\rho)e^{-\gamma c}\right], \tag{B.20b}$$

$$g = g_\infty - (g_\infty - g_0)e^{-\varepsilon c}, \tag{B.20c}$$

where $c = \alpha d$ is the light attenuation parameter along the diameter of a particle d, and $\alpha = 4\pi\chi/\lambda$, χ is the imaginary part of the refractive index of particles. The parameters ρ, ε, γ, g_0, g_∞ depend on the real part of the refractive index n of particles and are almost constant with the wavelength in the visible and near infrared for most materials. They are given in Table B.1 for various values of the real part of refractive index. It should be pointed out that the parameters listed in Table B.1 have a simple physical meaning:

$$g_0 = g(c = 0), g_\infty = g(c \to \infty), \tag{B.21}$$

$$\rho = 1 - C_{\text{sca}}(c \to \infty)/\pi a^2, \tag{B.22}$$

and the values ε and γ can be derived by matching Eq. (B.20) with Mie theory or geometrical optics calculations (say, as $c \to 0$). The analytical equations for the parameters ρ, g_∞ as functions of n are given by Kokhanovsky (2001). The parameterization of $g_0, g_\infty, \varepsilon, \gamma$, and ρ as functions of n are given by Kokhanovsky (2021). An important point is that Eqs. (B.20) are valid for randomly oriented convex particles as well (Kokhanovsky et al., 2024). The values of ρ, g_∞ remain the same as for spheres with the same real part of refractive index and the value of d is defined as: $d = 6V/S$, where V is the volume of a particle and S is its surface area. The parameters ε, γ, and g_0 depend on the shape of particles. In particular, Kokhanovsky et al. (2024) have proposed the following parameterizations for the fractal ice crystals:

$$g_\infty = 1.008 - 0.11(n-1), \tag{B.23}$$

Table B.1 The parameters in Eq. (B.20) (Kokhanovsky, 2001, 2021; Kokhanovsky et al., 2024).

n	ρ	g_0	g_∞	γ	ε
1.1	0.0252	0.9731	0.9946	0.7677	0.5180
1.2	0.0443	0.9341	0.9856	0.8348	0.6528
1.25	0.0529	0.9147	0.9806	0.8623	0.6948
1.3	0.0611	0.8961	0.9751	0.8869	0.7280
1.333	0.0664	0.8843	0.9714	0.9020	0.7468
1.34	0.0675	0.8818	0.9706	0.9051	0.7505
1.35	0.0691	0.8783	0.9695	0.9095	0.7555
1.4	0.0768	0.8613	0.9638	0.9302	0.7785
1.45	0.0844	0.8542	0.9579	0.9496	0.7985
1.5	0.0918	0.8299	0.9520	0.9677	0.8160
1.55	0.0991	0.8154	0.9460	0.9849	0.8315
1.6	0.1063	0.8015	0.9400	1.0012	0.8453

$$\rho = 0.0123 + 0.1622(n - 1), \tag{B.24}$$

$$g_0 = 0.9919 - 0.769(n - 1), \tag{B.25}$$

$\gamma = 0.8571$ and $\varepsilon = 0.9045$.

It should be pointed out that it follows from Eqs. (B.20) for the absorption cross-section $C_{abs} = C_{ext} - C_{sca}$ and the probability of photon absorption $\beta = C_{abs}/C_{ext}$:

$$C_{abs} = \pi a^2(1 - \rho)(1 - \exp(-\gamma c)), \tag{B.26}$$

$$\beta = \frac{1}{2}(1 - \rho)(1 - \exp(-\gamma c)). \tag{B.27}$$

Figures B.1–B.4 show spectral dependencies of the local optical parameters calculated using the approximation for fractal ice grains given earlier (Kokhanovsky et al., 2024). One can see that the spectral local optical properties of large scatterers depend on the value of the spectral attenuation parameter c shown in Figure B.5 for ice particles. The local optical parameters shown in Figures B.1–B.3 can be used to derive global optical properties of turbid media (e.g. spectral transmittance and reflectance). In particular, the spectral spherical albedo can be calculated using the following analytical equation valid for weakly absorbing semi-infinite media (Zege et al., 1991):

$$r = \exp(-4s), \tag{B.28}$$

where

$$s = \sqrt{\frac{1 - \omega_0}{3(1 - g)}} \tag{B.29}$$

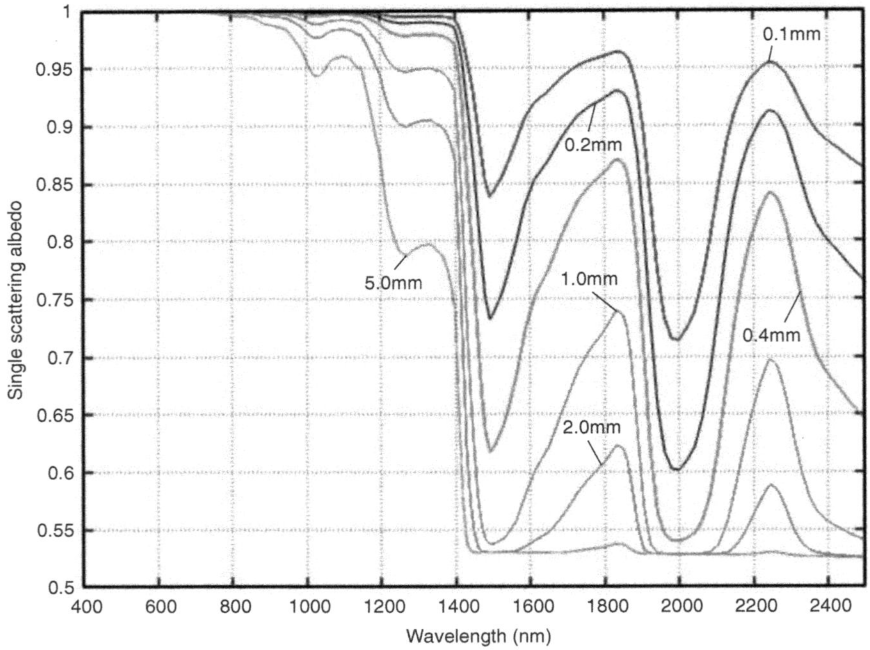

Figure B.1 The dependence of the single-scattering albedo on the wavelength for various effective diameters of snow grains. *Source*: Adapted from Kokhanovsky et al. (2024) (licensed under CC BY 4.0).

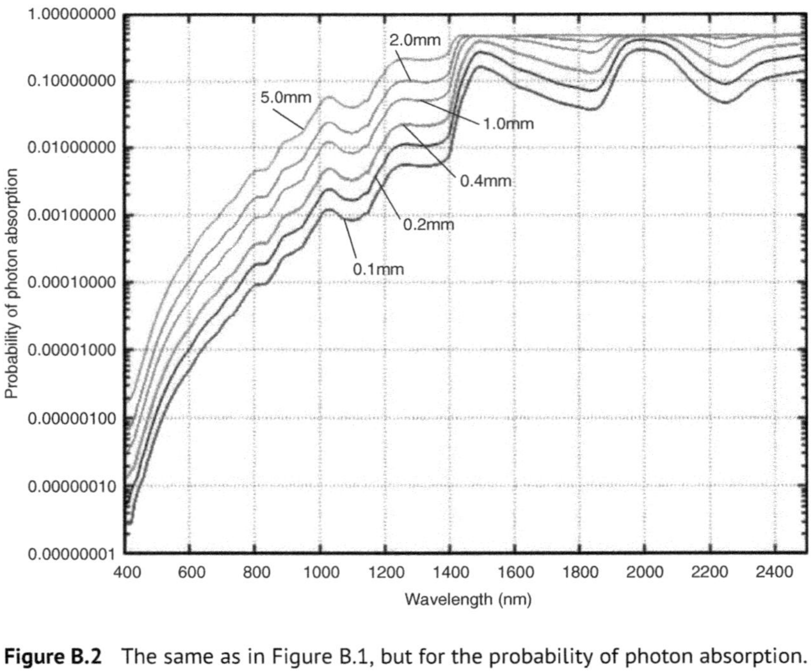

Figure B.2 The same as in Figure B.1, but for the probability of photon absorption.

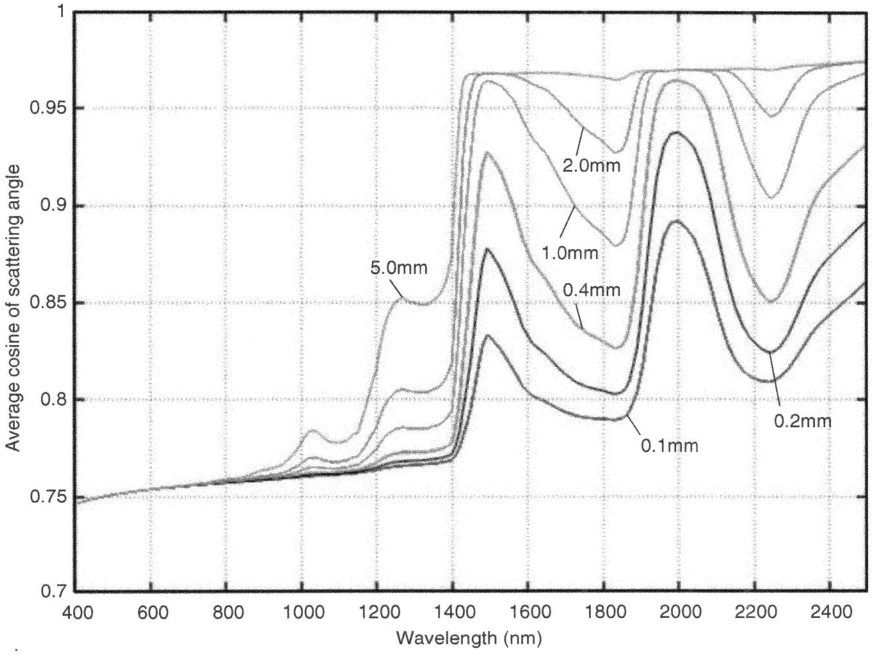

Figure B.3 The same as in Figure B.1, but for average cosine of scattering angle.

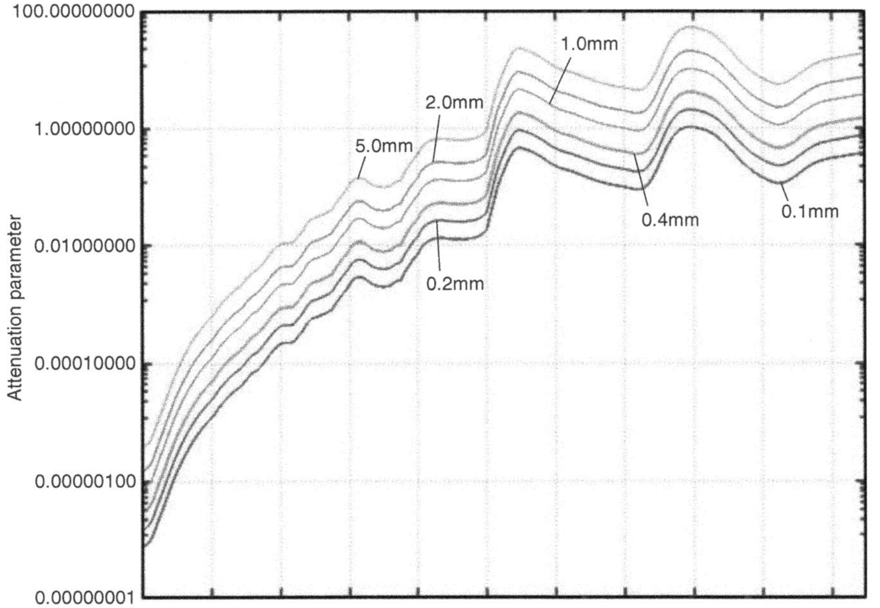

Figure B.4 The same as in Figure B.1, but for attenuation parameter.

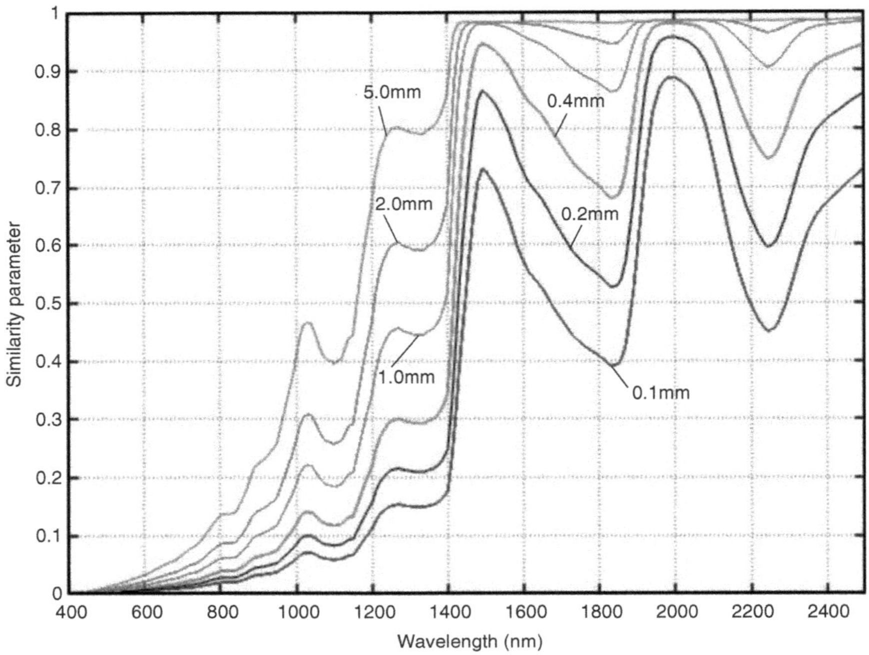

Figure B.5 The same as in Figure B.1, but for similarity parameter.

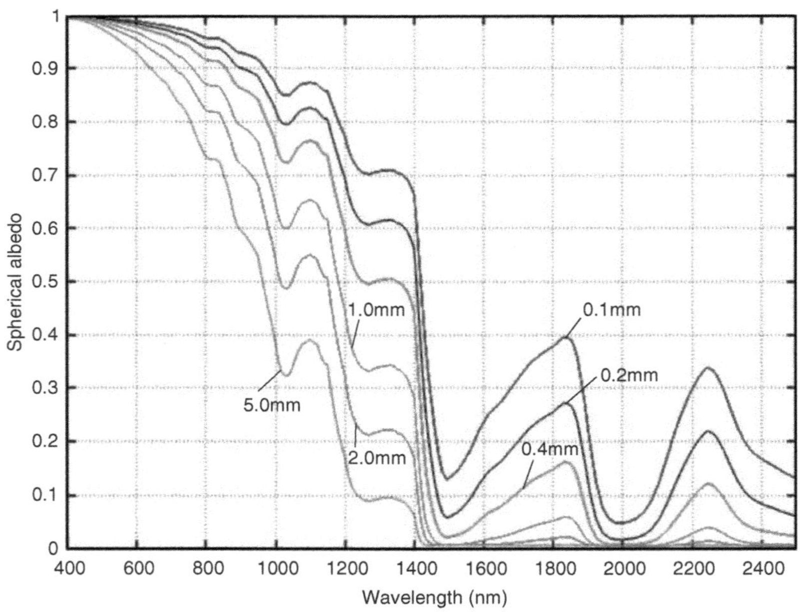

Figure B.6 The same as in Figure B.1, but for spherical albedo.

is the similarity parameter. The spectra of s and r calculated using Eqs. (B.28), (B.29), (B.20b), and (B.20c) are given in Figures B.5 and B.6, respectively. The upper curve in Figure B.6 corresponds to the fine-grained snow common in the interior of Antarctica. The lower curve corresponds to the case of melting wet snow with large grains.

Bibliography

V.P. Afanas'ev, D.S. Efremenko, and A.V. Lubenchenko. On the application of the invariant embedding method and the radiative transfer equation codes for surface state analysis. In *Light Scattering Reviews*, pages 363–423. Springer, Berlin, Heidelberg, 2013. ISBN 9783642321061.

J.B. Aizenberg and V.P. Budak. The science of light engineering, fields of application and theoretical foundations. *Light & Engineering*, 26:4–6, 2018. https://doi.org/10.33383/2018-045.

V.A. Ambartsumian. On a one-dimensional case of the problem of a scattering and absorbing medium of finite optical depth. *Communications of AN Armenian SSR, Natural Sciences Series*, 1(2):31–36, 1944a.

V.A. Ambartsumian. On the problem of diffuse reflection of light. *Journal of Physics of the USSR*, 8:65–75, 1944b.

V.A. Ambartsumyan. The scattering of light by planetary atmospheres. *Astronomicheskii Zhurnal*, 19: 30, 1942. [in Russian].

V. Ambartzumian. The effect of the absorption lines on the radiative equilibrium of the outer layers of the stars. *Publications of the Astronomical Observatory of Leningrad University*, 6:7–18, 1936.

L.A. Apresyan and Y.A. Kravtsov. *Radiation Transfer: Statistical and Wave Aspects*. Gordon and Breach, London, NY, 1996. ISBN 978-2881249204.

A. Arking and K. Grossman. The influence of line shape and band structure on temperatures in planetary atmospheres. *Journal of the Atmospheric Sciences*, 29(5):937–949, 1972. https://doi.org/10.1175/1520-0469(1972)029<0937:tiolsa>2.0.co;2.

W. Armbruster and J. Fischer. Improved method of exponential sum fitting of transmissions to describe the absorption of atmospheric gases. *Applied Optics*, 35(12): 1931, 1996. https://doi.org/10.1364/ao.35.001931.

K. Bai, K. Li, L. Shao, et al. LGHAP v2: a global gap-free aerosol optical depth and PM2.5 concentration dataset since 2000 derived via big earth data analytics. *Earth System Science Data*, 16(5):2425–2448, 2024. ISSN 1866-3516. https://doi.org/10.5194/essd-16-2425-2024.

Y.N. Barabanenkov, Y.A. Kravtsov, V.D. Ozrin, and A.I. Saichev. II enhanced backscattering in optics. In *Progress in Optics*, pages 65–197, Elsevier, 1991.

Analytical Methods in Radiative Transfer, First Edition. Alexander Kokhanovsky, Vijay Natraj and Dmitry Efremenko.

B.A. Baum, P. Yang, A.J. Heymsfield, et al. Ice cloud single-scattering property models with the full phase matrix at wavelengths from 0.2 to 100 m. *Journal of Quantitative Spectroscopy and Radiative Transfer*, 146:123–139, 2014. ISSN 0022-4073. https://doi.org/10.1016/j.jqsrt.2014.02.029.

E. Boesche, P. Stammes, R. Preusker, et al. Polarization of skylight in the O_2A band: effects of aerosol properties. *Applied Optics*, 47(19): 3467, 2008. https://doi.org/10.1364/ao.47.003467.

C.F. Bohren and D.R. Huffman. *Absorption and Scattering of Light by Small Particles*. Wiley, 1998. https://doi.org/10.1002/9783527618156.

L. Boltzmann. Ableitung des Stefan'en Gesetzes, betreffend die Abhängigkeit der Wärmestrahlung von der Temperatur aus der electromagnetischen Lichttheorie. *Annalen der Physik*, 258(6):291–294, 1884. https://doi.org/10.1002/andp.18842580616.

J. Brence, J. Tanevski, J. Adams, et al. Surrogate models of radiative transfer codes for atmospheric trace gas retrievals from satellite observations. *Machine Learning*, 112(4):1337–1363, 2022. ISSN 1573-0565. https://doi.org/10.1007/s10994-022-06155-2.

M. Buchwitz, V.V. Rozanov, and J.P. Burrows. A correlated-k distribution scheme for overlapping gases suitable for retrieval of atmospheric constituents from moderate resolution radiance measurements in the visible/near-infrared spectral region. *Journal of Geophysical Research: Atmospheres*, 105(D12):15247–15261, 2000. https://doi.org/10.1029/2000jd900171.

M.I. Budyko. The effect of solar radiation variations on the climate of the earth. *Tellus A: Dynamic Meteorology and Oceanography*, 21(5): 611, 1969. ISSN 1600-0870. https://doi.org/10.3402/tellusa.v21i5.10109.

B.D. Bue, D.R. Thompson, S. Deshpande, et al. Neural network radiative transfer for imaging spectroscopy. *Atmospheric Measurement Techniques*, 12(4):2567–2578, 2019. ISSN 1867-8548. https://doi.org/10.5194/amt-12-2567-2019.

Y. Cao, W. Zhang, Y. Zhang, et al. A new k-interval selection technique for fast atmospheric radiance calculation in remote sensing applications. *Journal of Quantitative Spectroscopy and Radiative Transfer*, 112(9):1479–1485, 2011. https://doi.org/10.1016/j.jqsrt.2011.03.004.

K.M. Case and P.F. Zweifel. *Linear Transport Theory*. Addison-Wesley Pub. Co., Reading, MA, 1967.

P. Ceppi, F. Brient, M.D. Zelinka, and D.L. Hartmann. Cloud feedback mechanisms and their representation in global climate models. *WIREs Climate Change*,8(4): e465, 2017. ISSN 1757-7799. https://doi.org/10.1002/wcc.465.

S. Chandrasekhar. *Radiative Transfer*. Dover Publications, Inc., New York, 1950.

H. Chen, V. V. Ginzburg, J. Yang, et al. Thermal conductivity of polymer-based composites: fundamentals and applications. *Progress in Polymer Science*, 59:41–85, 2016. ISSN 0079-6700. https://doi.org/10.1016/j.progpolymsci.2016.03.001.

J. Cheng, H. Wu, Z. Liu, et al. Vegetation feedback causes delayed ecosystem response to East Asian summer monsoon rainfall during the Holocene. *Nature Communications*, 12(1): 1843, 2021. ISSN 2041-1723. https://doi.org/10.1038/s41467-021-22087-2.

M.-D. Chou, W.L. Ridgway, and M.M.-H. Yan. One-parameter scaling and exponential-sum fitting for water vapor and CO_2 infrared transmission functions. *Journal of the Atmospheric Sciences*, 50(14):2294–2303, 1993. https://doi.org/10.1175/1520-0469(1993)050<2294:opsaes>2.0.co;2.

M.D. Chou and A. Arking. Computation of infrared cooling rates in the water vapor bands. *Journal of Atmospheric Sciences*, 37:855–867, 1980.

O.D. Chwolson. Grundzüge einer matimatischen Theorie der inneren Diffusion des Licht. *Bulletin de l'Académie Impériale des Sciences de Saint-Pétersbourg*, 33:221–256, 1889. [in German].

W.M. Cornette and J.G. Shanks. Physically reasonable analytic expression for the single-scattering phase function. *Applied Optics*, 31(16): 3152, 1992. https://doi.org/10.1364/ao.31.003152.

D. Crisp. Absorption of sunlight by water vapor in cloudy conditions: a partial explanation for the cloud absorption anomaly. *Geophysical Research Letters*, 24(5):571–574, 1997. https://doi.org/10.1029/97gl50245.

J.V. Dave and B.H. Armstrong. Computations of high-order associated legendre polynomials. *Journal of Quantitative Spectroscopy and Radiative Transfer*, 10(6):557–562, 1970. https://doi.org/10.1016/0022-4073(70)90073-7.

P. Debye. Der Lichtdruck auf Kugeln von beliebigem Material. *Annalen der Physik*, 335(11):57–136, 1909. https://doi.org/10.1002/andp.19093351103. [in German].

D. Deirmendjian. *Electromagnetic Scattering on Spherical Polydispersions*. Elsevier Science Ltd., New York, 1969.

A. del Águila and D.S. Efremenko. Fast hyper-spectral radiative transfer model based on the double cluster low-streams regression method. *Remote Sensing*, 13(3): 434, 2021. ISSN 2072-4292. https://doi.org/10.3390/rs13030434.

A. del Águila, D. Efremenko, V. Molina García, and J. Xu. Analysis of two dimensionality reduction techniques for fast simulation of the spectral radiances in the Hartley-Huggins band. *Atmosphere*, 10(3): 142, 2019a. ISSN 2073-4433. https://doi.org/10.3390/atmos10030142.

A. del Águila, D.S. Efremenko, and T. Trautmann. A review of dimensionality reduction techniques for processing hyper-spectral optical signal. *Light and Engineering*, 27(3):85–98, 2019b. https://doi.org/10.33383/2019-017.

A. del Águila, D.S. Efremenko, V. Molina García, and M.Y. Kataev. Cluster low-streams regression method for hyperspectral radiative transfer computations: cases of O_2A- and CO_2 bands. *Remote Sensing*, 12(8): 1250, 2020. ISSN 2072-4292. https://doi.org/10.3390/rs12081250.

A. Doicu and M.I. Mishchenko. Overview of methods for deriving the radiative transfer theory from the Maxwell equations. I: approach based on the far-field Foldy equations. *Journal of Quantitative Spectroscopy and Radiative Transfer*, 220:123–139, 2018. https://doi.org/10.1016/j.jqsrt.2018.09.004.

A. Doicu and M.I. Mishchenko. An overview of methods for deriving the radiative transfer theory from the Maxwell equations. II: approach based on the Dyson and Bethe-Salpeter equations. *Journal of Quantitative Spectroscopy and Radiative Transfer*, 224:25–36, 2019. https://doi.org/10.1016/j.jqsrt.2018.10.032.

A. Doicu, T. Trautmann, and F. Schreier. *Numerical Regularization for Atmospheric Inverse Problems*. Springer, Berlin, Heidelberg, 2010. ISBN 9783642054396. https://doi.org/10.1007/978-3-642-05439-6.

A. Doicu, A. Doicu, D. Efremenko, and T. Trautmann. Cloud tomographic retrieval algorithms. I: surrogate minimization method. *Journal of Quantitative Spectroscopy and Radiative Transfer*, 277: 107954, 2022. ISSN 0022-4073. https://doi.org/10.1016/j.jqsrt.2021.107954.

L.S. Dolin. Scattering of light in a layer of turbid medium. *Izvestiya Vysshikh Uchebnykh Zavedenii. Radiofizika*, 7(2):380–382, 1964.

L.A. Dombrovsky and D. Baillis. *Thermal Radiation in Disperse Systems: An Engineering Approach*. Begell House Publishers Inc., U.S., 2010.

M. Duan. A fast radiative transfer model for simulating high-resolution absorption bands. *Journal of Geophysical Research*, 110:D15201, 2005. https://doi.org/10.1029/2004jd005590.

O. Dubovik, B. Holben, T.F. Eck, et al. Variability of absorption and optical properties of key aerosol types observed in worldwide locations. *Journal of the Atmospheric Sciences*, 59(3):590–608, 2002. ISSN 1520-0469. https://doi.org/10.1175/1520-0469(2002)059<0590:voaaop>2.0.co;2.

O. Dubovik, M. Herman, A. Holdak, et al. Statistically optimized inversion algorithm for enhanced retrieval of aerosol properties from spectral multi-angle polarimetric satellite observations. *Atmospheric Measurement Techniques*, 4(5):975–1018, May 2011. ISSN 1867-8548. https://doi.org/10.5194/amt-4-975-2011.

D. Efremenko and A. Kokhanovsky. *Foundations of Atmospheric Remote Sensing*. Springer International Publishing, 2021. https://doi.org/10.1007/978-3-030-66745-0.

D. Efremenko, A. Doicu, D. Loyola, and T. Trautmann. Optical property dimensionality reduction techniques for accelerated radiative transfer performance: application to remote sensing total ozone retrievals. *Journal of Quantitative Spectroscopy and Radiative Transfer*, 133:128–135, 2014. ISSN 0022-4073. https://doi.org/10.1016/j.jqsrt.2013.07.023.

D. Efremenko, B. Pflug, R. Richter, et al. Fast computations of the top-of-the-atmosphere radiance in a spectral range 400–2500 nm using the PYDOME tool. In *The 5th International Electronic Conference on Remote Sensing*, ECRS 2023. MDPI, 2023. https://doi.org/10.3390/ecrs2023-15858.

D.S. Efremenko. Discrete ordinate radiative transfer model with the neural network based eigenvalue solver: proof of concept. *Light and Engineering*, 1:56–62, 2021. ISSN 1068-9761. https://doi.org/10.33383/2020-075.

D.S. Efremenko, D.G. Loyola R., P. Hedelt, and R.J.D. Spurr. Volcanic SO_2 plume height retrieval from UV sensors using a full-physics inverse learning machine algorithm. *International Journal of Remote Sensing*, 38(sup1):1–27, 2017a. ISSN 1366-5901. https://doi.org/10.1080/01431161.2017.1348644.

D.S. Efremenko, V. Molina García, S. Gimeno García, and A. Doicu. A review of the matrix-exponential formalism in radiative transfer. *Journal of Quantitative Spectroscopy and Radiative Transfer*, 196:17–45, 2017b. https://doi.org/10.1016/j.jqsrt.2017.02.015.

P. Eriksson, R. Ekelund, J. Mendrok, et al. A general database of hydrometeor single scattering properties at microwave and sub-millimetre wavelengths. *Earth System Science Data*, 10(3):1301–1326, 2018. ISSN 1866-3516. https://doi.org/10.5194/essd-10-1301-2018.

M.G. Flanner, C.S. Zender, J.T. Randerson, and P.J. Rasch. Present-day climate forcing and response from black carbon in snow. *Journal of Geophysical Research: Atmospheres*, 112:D11202, 2007. ISSN 0148-0227. https://doi.org/10.1029/2006jd008003.

K. Fodor. *A Survey of Dimension Reduction Techniques*. Lawrence Livermore National Lab. (LLNL), Livermore, CA, 2002.

V.A. Fok. On some integral equations of mathematical physics. *Mathematical Collection*, 14(56):3–50, 1944. [in Russian].

L.L. Foldy. The multiple scattering of waves. I. General theory of isotropic scattering by randomly distributed scatterers. *Physical Review*, 67(3-4):107–119, 1945. https://doi.org/10.1103/physrev.67.107.

J.R. Frisvad and H. Kragh. On Ludvig Lorenz and his 1890 treatise on light scattering by spheres. *The European Physical Journal H*, 44(2):137–160, 2019. ISSN 2102-6467. https://doi.org/10.1140/epjh/e2019-100022-y.

Q. Fu and K.N. Liou. On the correlatedk-distribution method for radiative transfer in nonhomogeneous atmospheres. *Journal of the Atmospheric Sciences*, 49(22):2139–2156, 1992. https://doi.org/10.1175/1520-0469(1992)049<2139:otcdmf>2.0.co;2.

C. Gao, D. Liang, B. Sun, et al. Linearized single-scattering property database for hexagonal prism ice particles. *Remote Sensing*, 14(23): 6138, December 2022. ISSN 2072-4292. https://doi.org/10.3390/rs14236138.

R. Goody, R. West, L. Chen, and D. Crisp. The correlated-k method for radiation calculations in nonhomogeneous atmospheres. *Journal of Quantitative Spectroscopy and Radiative Transfer*, 42(6):539–550, 1989. https://doi.org/10.1016/0022-4073(89)90044-7.

A.N. Gorban, B. Kégl, D.C. Wunsch, and A.Y. Zinovyev. editors. *Principal Manifolds for Data Visualization and Dimension Reduction*. Springer, Berlin, Heidelberg, 2008.

A. Hall and S. Manabe. The role of water vapor feedback in unperturbed climate variability and global warming. *Journal of Climate*, 12(8):2327–2346, 1999. ISSN 1520-0442. https://doi.org/10.1175/1520-0442(1999)012<2327:trowvf>2.0.co;2.

J. Hansen and L. Nazarenko. Soot climate forcing via snow and ice albedos. *Proceedings of the National Academy of Sciences*, 101(2):423–428, 2003. ISSN 1091-6490. https://doi.org/10.1073/pnas.2237157100.

J. Hansen, D. Johnson, A. Lacis, et al. Climate impact of increasing atmospheric carbon dioxide. *Science*, 213(4511):957–966, 1981. ISSN 1095-9203. https://doi.org/10.1126/science.213.4511.957.

J. E. Hansen. Exact and approximate solutions for multiple scattering by cloudy and hazy planetary atmospheres. *Journal of the Atmospheric Sciences*, 26(3):478–487, 1969. https://doi.org/10.1175/1520-0469(1969)026<0478:eaasfm>2.0.co;2.

J.E. Hansen and L.D. Travis. Light scattering in planetary atmospheres. *Space Science Reviews*, 16(4):527–610, 1974. https://doi.org/10.1007/bf00168069.

B. Hapke. *Theory of Reflectance and Emittance Spectroscopy*. Cambridge University Press, New York, 1993.

B. Hapke. *Theory of Reflectance and Emittance Spectroscopy*. Cambridge University Press, 2009. https://doi.org/10.1017/cbo9781139025683.

O.P. Hasekamp and A. Butz. Efficient calculation of intensity and polarization spectra in vertically inhomogeneous scattering and absorbing atmospheres. *Journal of Geophysical Research*, 113:D20309, 2008. https://doi.org/10.1029/2008jd010379.

L.G. Henyey and J.L. Greenstein. Diffuse radiation in the galaxy. *The Astrophysical Journal*, 93:70–83, 1941.

M. Hess, P. Koepke, and I. Schult. Optical properties of aerosols and clouds: the software package OPAC. *Bulletin of the American Meteorological Society*, 79(5):831–844, 1998. https://doi.org/10.1175/1520-0477(1998)079<0831:opoaac>2.0.co;2.

A. Hollstein and R. Lindstrot. Fast reconstruction of hyperspectral radiative transfer simulations by using small spectral subsets: application to the oxygen A band. *Atmospheric Measurement Techniques*, 7(2):599–607, 2014. https://doi.org/10.5194/amt-7-599-2014.

J.W. Hovenier, C. Van Der Mee, and H. Domke. *Transfer of Polarized Light in Planetary Atmospheres*. Springer, Netherlands, 2004.

Y.-X. Hu, B. Wielicki, B. Lin, et al. δ-fit: a fast and accurate treatment of particle scattering phase functions with weighted singular-value decomposition least-squares fitting. *Journal of Quantitative Spectroscopy and Radiative Transfer*, 65(4):681–690, 2000. ISSN 0022-4073. https://doi.org/10.1016/s0022-4073(99)00147-8.

J. Huang, J. Guo, F. Wang, et al. CALIPSO inferred most probable heights of global dust and smoke layers. *Journal of Geophysical Research: Atmospheres*, 120(10):5085–5100, 2015. ISSN 2169-8996. https://doi.org/10.1002/2014jd022898.

G.E. Hunt and I.P. Grant. Discrete space theory of radiative transfer and its application to problems in planetary atmospheres. *Journal of the Atmospheric Sciences*, 26(5):963–972, 1969. https://doi.org/10.1175/1520-0469(1969)026<0963:dstort>2.0.co;2.

G.E. Hunt. A review of computational techniques for analysing the transfer of radiation through a model cloudy atmosphere. *Journal of Quantitative Spectroscopy and Radiative Transfer*, 11(6):655–690, 1971. https://doi.org/10.1016/0022-4073(71)90046-x.

P.D. Hurley, S. Oliver, D. Farrah, et al. Principal component analysis and radiative transfer modelling of Spitzer Infrared spectrograph spectra of ultraluminous infrared galaxies. *Monthly Notices of the Royal Astronomical Society*, 424(3):2069–2078, 2012. https://doi.org/10.1111/j.1365-2966.2012.21352.x.

A. Ishimaru. *Wave Propagation and Scattering in Random Media*. Elsevier, 1978.

L.K. Jänicke, R. Preusker, N. Docter, and J. Fischer. Estimation of aerosol layer height from OLCI measurements in the O2A-absorption band over oceans. *Remote Sensing*, 15(16): 4080, 2023. ISSN 2072-4292. https://doi.org/10.3390/rs15164080.

J. Jendersie and E. d'Eon. An approximate Mie scattering function for fog and cloud rendering. In *ACM SIGGRAPH 2023 Talks*, SIGGRAPH'23. ACM, 2023. https://doi.org/10.1145/3587421.3595409.

H. Kagiwada and R. Kalaba. RAND Report RM-5537-PR. RAND Corporation, Santa Monica, CA, 1967.

E. Kamke. *Differentialgleichungen. Lösungsmethoden und Lösungen, Part 1, Gewöhnliche Differentialgleichungen*. Leipzig, 1959.

I.L. Katsev, A.S. Prikhach, E.P. Zege, et al. Speeding up the aerosol optical thickness retrieval using analytical solutions of radiative transfer theory. *Atmospheric Measurement Techniques*, 3(5):1403–1422, 2010. ISSN 1867-8548. https://doi.org/10.5194/amt-3-1403-2010.

G.W. Kattawar, A.T. Young, and T.J. Humphreys. Inelastic scattering in planetary atmospheres. I. The ring effect, without aerosols. *The Astrophysical Journal*, 243:1049–1057, 1981. https://doi.org/10.1086/158669.

K. Kawabata and S. Ueno. The first three orders of scattering in vertically inhomogeneous scattering-absorbing media. *Astrophysics and Space Science*, 150(2):327–344, 1988. ISSN 1572-946X. https://doi.org/10.1007/bf00641728.

J.R. Key and A.J. Schweiger. Tools for atmospheric radiative transfer: streamer and FluxNet. *Computers & Geosciences*, 24(5):443–451, 1998. https://doi.org/10.1016/s0098-3004(97)00130-1.

G. Kirchhoff and R. Bunsen. Chemische analyse durch spectralbeobachtungen. *Annalen der Physik und Chemie*, 189(7):337–381, 1861. https://doi.org/10.1002/andp.18611890702.

A. Kokhanovsky. Photon transport in asymmetric random media. *Journal of Optics A: Pure and Applied Optics*, 4(5):521–526, 2002. ISSN 1464-4258. https://doi.org/10.1088/1464-4258/4/5/306.

A. Kokhanovsky. *Snow Optics*. Springer International Publishing, 2021. ISBN 9783030865894. https://doi.org/10.1007/978-3-030-86589-4.

A. Kokhanovsky, M. Lamare, O. Danne, et al. Retrieval of snow properties from the Sentinel-3 ocean and land colour instrument. *Remote Sensing*, 11(19): 2280, 2019. ISSN 2072-4292. https://doi.org/10.3390/rs11192280.

A. Kokhanovsky, M. Brell, K. Segl, and S. Chabrillat. SNOWTRAN: a fast radiative transfer model for polar hyperspectral remote sensing applications. *Remote Sensing*, 16(2): 334, 2024. ISSN 2072-4292. https://doi.org/10.3390/rs16020334.

A.A. Kokhanovsky. *Optics of Light Scattering Media*. Springer-Praxis, Chichester, 2001.

A.A. Kokhanovsky. Small-angle approximations of the radiative transfer theory. *Journal of Physics D: Applied Physics*, 30(20):2837–2840, 1997. https://doi.org/10.1088/0022-3727/30/20/009.

A. Kokhanovsky. *Cloud Optics*. Springer, Netherlands, 2006. ISBN 9781402040207.

A.A. Kokhanovsky, A.B. Davis, B. Cairns, et al. Space-based remote sensing of atmospheric aerosols: the multi-angle spectro-polarimetric frontier. *Earth-Science Reviews*, 145:85–116, 2015. ISSN 0012-8252. https://doi.org/10.1016/j.earscirev.2015.01.012.

S. Korkin, A.M. Sayer, A. Ibrahim, and A. Lyapustin. A practical guide to writing a radiative transfer code. *Computer Physics Communications*, 271: 108198, 2022. https://doi.org/10.1016/j.cpc.2021.108198.

V. Kourganoff. *Basic Methods in Transfer Problems; Radiative Equilibrium and Neutron Diffusion*. Clarendon Press, Oxford, 1952.

D.P. Kratz. The correlated k-distribution technique as applied to the AVHRR channels. *Journal of Quantitative Spectroscopy and Radiative Transfer*, 53(5):501–517, 1995. https://doi.org/10.1016/0022-4073(95)90050-0.

D.P. Kratz, M.-D. Chou, M.M.-H. Yan, and C.-H. Ho. Minor trace gas radiative forcing calculations using the k distribution method with one-parameter scaling. *Journal of Geophysical Research: Atmospheres*, 103(D24):31647–31656, 1998. https://doi.org/10.1029/1998jd200009.

Y.A. Kravtsov and L.A. Apresyan. IV radiative transfer: new aspects of the old theory. In *Progress in Optics*, pages 179–244. Elsevier, 1996. https://doi.org/10.1016/s0079-6638(08)70315-9.

I. Kuščer. Milne's problem for anisotropic scattering. *Journal of Mathematics and Physics*, 34 (1–4):256–266, 1955. ISSN 0097-1421. https://doi.org/10.1002/sapm1955341256.

A. Lacis, W.C. Wang, and J. Hansen. Correlated k-distribution method for radiative transfer in climate models: application to effect of cirrus clouds on climate. *NASA Conference Publication* 2076:309–314, 1979.

A.A. Lacis and V. Oinas. A description of the correlated k distribution method for modeling nongray gaseous absorption, thermal emission, and multiple scattering in vertically inhomogeneous atmospheres. *Journal of Geophysical Research*, 96(D5):9027, 1991. https://doi.org/10.1029/90jd01945.

I.E. Lagaris, A. Likas, and D.I. Fotiadis. Artificial neural networks for solving ordinary and partial differential equations. *IEEE Transactions on Neural Networks*, 9(5):987–1000, 1998. ISSN 1045-9227. https://doi.org/10.1109/72.712178.

J.H. Lambert. *Photometria, sive de mensura et gradibus luminis, colorum et umbrae*. Klett, Augustae Vindelicorum, 1760.

M. Lax. Multiple scattering of waves. *Reviews of Modern Physics*, 23(4):287–310, 1951. https://doi.org/10.1103/revmodphys.23.287.

P.F. Levelt, G.H.J. van den Oord, M.R. Dobber, et al. The ozone monitoring instrument. *IEEE Transactions on Geoscience and Remote Sensing*, 44(5):1093–1101, 2006. https://doi.org/10.1109/tgrs.2006.872333.

C. Li, J. Joiner, N.A. Krotkov, and P.K. Bhartia. A fast and sensitive new satellite SO_2 retrieval algorithm based on principal component analysis: application to the ozone monitoring instrument. *Geophysical Research Letters*, 40(23):6314–6318, 2013. ISSN 1944-8007.

Z. Lin, N. Chen, Y. Fan, et al. New treatment of strongly anisotropic scattering phase functions: the delta-$M+$ method. *Journal of the Atmospheric Sciences*, 75(1):327–336, 2018. https://doi.org/10.1175/jas-d-17-0233.1.

K.N. Liou. *An Introduction to Atmospheric Radiation*. Academic Press, San Diego, 2002a.

K.N. Liou. *An Introduction to Atmospheric Radiation, 2nd Edition*. Academic Press, 2002b.

P. Liu. A survey of remote-sensing big data. *Frontiers in Environmental Science*, 3: 45, 2015.

X. Liu, W.L. Smith, D.K. Zhou, and A. Larar. Principal component-based radiative transfer model for hyperspectral sensors: theoretical concept. *Applied Optics*, 45(1): 201, 2006. https://doi.org/10.1364/ao.45.000201.

E. Lommel. Die Photometric der diffusen Zuruckwerfung. *Sitzungsberichte der Akademie der Wissenschaften zu München*, 17:95–124, 1887. [in German].

L.V. Lorentz. Lysbevægelser i og uden for en af plane Lysbølger belyst Kugle. *Det kongelige danske Videnskabernes Selskabs Skrifter*, 6(6):1–62, 1890.

D.G. Loyola, S. Gimeno García, R. Lutz, et al. The operational cloud retrieval algorithms from TROPOMI on board Sentinel-5 precursor. *Atmospheric Measurement Techniques*, 11(1):409–427, 2018. ISSN 1867-8548. https://doi.org/10.5194/amt-11-409-2018.

Y. Ma, H. Wu, L. Wang, et al. Remote sensing big data computing: challenges and opportunities. *Future Generation Computer Systems*, 51:47–60, 2015.

R.H. MacArthur. On the relative abundance of bird species. *Proceedings of the National Academy of Sciences of the United States of America*, 43(3):293–295, 1957.

A. Macke, J. Mueller, and E. Raschke. Single scattering properties of atmospheric ice crystals. *Journal of the Atmospheric Sciences*, 53(19):2813–2825, 1996. https://doi.org/10.1175/1520-0469(1996)053<2813:sspoai>2.0.co;2.

Y. Mano. Modified ESFT method for application to atmospheric radiation. *Papers in Meteorology and Geophysics*, 46(1):1–8, 1995. https://doi.org/10.2467/mripapers.46.1.

Mathieu and Edouard Roser. Max Data page: Annual number of objects launched into space. Data adapted from United Nations Office for Outer Space Affairs, 2022. URL https://ourworldindata.org/grapher/yearly-number-of-objects-launched-into-outer-space.

M. Matricardi. A principal component based version of the RTTOV fast radiative transfer model. *Quarterly Journal of the Royal Meteorological Society*, 136:1823–1835, 2010.

V.S. Meadows and D. Crisp. Ground-based near-infrared observations of the Venus nightside: the thermal structure and water abundance near the surface. *Journal of Geophysical Research: Planets*, 101(E2):4595–4622, 1996. https://doi.org/10.1029/95je03567.

L. Mei, V. Rozanov, and J.P. Burrows. A fast and accurate radiative transfer model for aerosol remote sensing. *Journal of Quantitative Spectroscopy and Radiative Transfer*, 256: 107270, 2020.ISSN 0022-4073. https://doi.org/10.1016/j.jqsrt.2020.107270.

Z. Meng, P. Yang, G.W. Kattawar, et al. Single-scattering properties of tri-axial ellipsoidal mineral dust aerosols: a database for application to radiative transfer calculations. *Journal of Aerosol Science*, 41(5):501–512, 2010. ISSN 0021-8502. https://doi.org/10.1016/j.jaerosci.2010.02.008.

G. Mie. Beiträge zur optik trüber medien, speziell kolloidaler metallösungen. *Annalen der Physik*, 330(3):377–445, 1908. https://doi.org/10.1002/andp.19083300302. [in German].

E.A. Milne. Radiative equilibrium in the outer layers of a star: the temperature distribution and the law of darkening. *Monthly Notices of the Royal Astronomical Society*, 81(5):361–375, 1921. ISSN 1365-2966. https://doi.org/10.1093/mnras/81.5.361.

Q. Min and M. Duan. A successive order of scattering model for solving vector radiative transfer in the atmosphere. *Journal of Quantitative Spectroscopy and Radiative Transfer*, 87(3-4):243–259, 2004. https://doi.org/10.1016/j.jqsrt.2003.12.019.

M.I. Mishchenko, I.V. Geogdzhayev, W.B. Rossow, et al. Long-term satellite record reveals likely recent aerosol trend. *Science*, 315(5818):1543–1543, 2007. https://doi.org/10.1126/science.1136709.

M.I. Mishchenko. Multiple scattering by particles embedded in an absorbing medium. 1. Foldy–Lax equations, order-of-scattering expansion, and coherent field. *Optics Express*, 16(3): 2288, 2008. https://doi.org/10.1364/oe.16.002288.

M.I. Mishchenko, R.F. Cahalan and J. Fischer. editors. *Radiation Processes in the Atmosphere and Ocean (IRS2012): Proceedings of the International Radiation Symposium (IRC/IAMAS)*, 6–10 August 2012, Berlin, volume 1531 of AIP Conference Proceedings, 11–18, AIP, Melville, NY, 2013.

M.I. Mishchenko. *Electromagnetic Scattering by Particles and Particle Groups: An Introduction*. Cambridge, 2014.

M.I. Mishchenko, J.M. Dlugach, E.G. Yanovitskij, and N.T. Zakharova. Bidirectional reflectance of flat, optically thick particulate layers: an efficient radiative transfer solution and applications to snow and soil surfaces. *Journal of Quantitative Spectroscopy and Radiative Transfer*, 63 (2–6):409–432, 1999. https://doi.org/10.1016/s0022-4073(99)00028-x.

S. Mishra and R. Molinaro. Physics informed neural networks for simulating radiative transfer. *Journal of Quantitative Spectroscopy and Radiative Transfer*, 270: 107705, 2021.ISSN 0022-4073. https://doi.org/10.1016/j.jqsrt.2021.107705.

C. Mitrescu and G.L. Stephens. On similarity and scaling of the radiative transfer equation. *Journal of Quantitative Spectroscopy and Radiative Transfer*, 86(4):387–394, 2004. https://doi.org/10.1016/j.jqsrt.2003.12.028.

J.-L. Moncet, G. Uymin, A.E. Lipton, and H.E. Snell. Infrared radiance modeling by optimal spectral sampling. *Journal of the Atmospheric Sciences*, 65(12):3917–3934, 2008. https://doi.org/10.1175/2008jas2711.1.

P. Mouroulis, B. Van Gorp, R.O. Green, et al. Portable remote imaging spectrometer coastal ocean sensor: design, characteristics, and first flight results. *Applied Optics*, 53(7): 1363, 2014.ISSN 2155-3165. https://doi.org/10.1364/ao.53.001363.

R. Mueller, J. Trentmann, C. Träger-Chatterjee, et al. The role of the effective cloud albedo for climate monitoring and analysis. *Remote Sensing*, 3(11):2305–2320, 2011. ISSN 2072-4292. https://doi.org/10.3390/rs3112305.

O. Munoz, F. Moreno, D. Guirado, et al. The Amsterdam–Granada light scattering database. *Journal of Quantitative Spectroscopy and Radiative Transfer*, 113(7):565–574, May 2012. ISSN 0022-4073. https://doi.org/10.1016/j.jqsrt.2012.01.014.

D.I. Nagirner. Viktor V. Sobolev and radiative transfer theory. *Journal of Quantitative Spectroscopy and Radiative Transfer*, 183:4–37, 2016. ISSN 0022-4073. https://doi.org/10.1016/j.jqsrt.2016.06.010.

D.I. Nagirner and V.V. Ivanov. Chandrasekhar's H-function revisited. *Journal of Quantitative Spectroscopy and Radiative Transfer*, 246: 106914, 2020.ISSN 0022-4073. https://doi.org/10.1016/j.jqsrt.2020.106914.

V. Natraj, X. Jiang, R.L. Shia, et al. Application of principal component analysis to high spectral resolution radiative transfer: a case study of the band. *Journal of Quantitative Spectroscopy and Radiative Transfer*, 95(4):539–556, 2005. https://doi.org/10.1016/j.jqsrt.2004.12.024.

V. Natraj, R.-L. Shia, and Y.L. Yung. On the use of principal component analysis to speed up radiative transfer calculations. *Journal of Quantitative Spectroscopy and Radiative Transfer*, 111(5):810–816, 2010. https://doi.org/10.1016/j.jqsrt.2009.11.004.

A.G. Nikoghossian. Ambartsumian's invariance principle and some nonlinear relations in radiative transfer theory. *Astrophysics*, 52(3):431–439, 2009. ISSN 1573-8191. https://doi.org/10.1007/s10511-009-9079-z.

C.W. O'Dell. Acceleration of multiple-scattering, hyperspectral radiative transfer calculations via low-streams interpolation. *Journal of Geophysical Research*, 115:D10206, 2010. https://doi.org/10.1029/2009jd012803.

R. Patgiri and A. Ahmed. Big Data: The V's of the game changer paradigm. In *2016 IEEE 18th International Conference on High Performance Computing and Communications; IEEE 14th International Conference on Smart City; IEEE 2nd International Conference on Data Science and Systems (HPCC/SmartCity/DSS)*. IEEE, 2016. https://doi.org/10.1109/hpcc-smartcity-dss.2016.0014.

K. Pearson. LIII. On lines and planes of closest fit to systems of points in space. *The London, Edinburgh, and Dublin Philosophical Magazine and Journal of Science*, 2(11):559–572, 1901. ISSN 1941-5990. https://doi.org/10.1080/14786440109462720.

A. Peraiah. *An Introduction to Radiative Transfer: Methods and Applications in Astrophysics*. Cambridge University Press, 2001. ISBN 0521779898.

L.M. Perliski and S. Solomon. On the evaluation of air mass factors for atmospheric near-ultraviolet and visible absorption spectroscopy. *Journal of Geophysical Research*, 98(D6):10363, 1993. https://doi.org/10.1029/93jd00465.

U. Platt. Differential optical absorption spectroscopy (DOAS). *Analytical Chemistry Series*, 127:27–83, 1994.

U. Platt and J. Stutz. *Differential Optical Absorption Spectroscopy: Principles and Applications*. Springer-Verlag, Berlin, Heidelberg, 2008.

J.F. Potter. The delta function approximation in radiative transfer theory. *Journal of the Atmospheric Sciences*, 27(6):943–949, 1970. https://doi.org/10.1175/1520-0469(1970)027<0943:tdfair>2.0.co;2.

R. Richter. A fast atmospheric correction algorithm applied to Landsat TM images. *International Journal of Remote Sensing*, 11(1):159–166, 1990. ISSN 1366-5901. https://doi.org/10.1080/01431169008955008.

C.D. Rodgers. *Inverse Methods for Atmospheric Sounding: Theory and Practice*. World Scientific, 2000. ISBN 9789812813718. https://doi.org/10.1142/3171.

J. Rosen. Redundancy and superfluity for electromagnetic fields and potentials. *American Journal of Physics*, 48(12):1071–1073, 1980. ISSN 1943-2909. https://doi.org/10.1119/1.12289.

G.B. Rybicki and A.P. Lightman. *Radiative Processes in Astrophysics*. Wiley, 1985. ISBN 9783527618170. https://doi.org/10.1002/9783527618170.

M Saito, P. Yang, J. Ding, and X. Liu. A comprehensive database of the optical properties of irregular aerosol particles for radiative transfer simulations. *Journal of the Atmospheric Sciences*, 78:2089–2111, 2021. ISSN 1520-0469. https://doi.org/10.1175/jas-d-20-0338.1.

A. Schuster. Radiation through a foggy atmosphere. *The Astrophysical Journal*, 21: 1, 1905. https://doi.org/10.1086/141186.

E.A.G. Schuur, A.D. McGuire, C. Schädel, et al. Climate change and the permafrost carbon feedback. *Nature*, 520(7546):171–179, 2015. ISSN 1476-4687. https://doi.org/10.1038/nature14338.

H. Schwander, A. Kaifel, A. Ruggaber, and P. Koepke. Spectral radiative-transfer modeling with minimized computation time by use of a neural-network technique. *Applied Optics*, 40(3): 331, 2001. https://doi.org/10.1364/ao.40.000331.

K. Schwarzschild. Über das Gleichgewicht der Sonnenatmosphäre. *Nachrichten der Königlichen Gesellschaft der Wissenschaften zu Gottingen*, 1:41–53, 1906. [in German].

J. Slusser, K. Hammond, A. Kylling, et al. Comparison of air mass computations. *Journal of Geophysical Research: Atmospheres*, 101(D5):9315–9321, 1996. https://doi.org/10.1029/96jd00054.

V.V. Sobolev. *Light Scattering in Planetary Atmospheres*. Pergamon Press, 1975.

V. Sobolev. The approximative solution of the task about light scattering in the media with an arbitrary scattering indicatrix. *Astronomicheskii Zhurnal*, 20 (5–6): 14, 1943. [in Russian].

V.V. Sobolev. *The Transfer of Radiant Energy in the Atmospheres of Stars and Planets*. GITTL, Moscow, 1956. [in Russian].

R. Spurr, V. Natraj, C. Lerot, et al. Linearization of the principal component analysis method for radiative transfer acceleration: application to retrieval algorithms and sensitivity studies. *Journal of Quantitative Spectroscopy and Radiative Transfer*, 125:1–17, 2013. ISSN 0022-4073. https://doi.org/10.1016/j.jqsrt.2013.04.002.

K. Stamnes, S.C. Tsay, and T. Nakajima. Computation of eigenvalues and eigenvectors for the discrete ordinate and matrix operator methods in radiative transfer. *Journal of Quantitative Spectroscopy and Radiative Transfer*, 39(5):415–419, 1988. https://doi.org/10.1016/0022-4073(88)90107-0.

K. Stamnes, G.E. Thomas, and J.J. Stamnes. *Radiative Transfer in the Atmosphere and Ocean*. Cambridge University Press, 2017. https://doi.org/10.1017/9781316148549.

J. Stefan. Ueber die Beziehung zwischen der Waermestrahlung und der Temperatur. *Sitzungsberichte der Kaiserlichen Akademie der Wissenschaften in Wien*, 79:391–428, 1879. Aus der k.k. Hof-und Staatsdruckerei.

G.G. Stokes. On the intensity of the light reflected from or transmitted through a pile of plates. *Proceedings of the Royal Society of London*, 11:545–556, 1862. ISSN 2053-9126. https://doi.org/10.1098/rspl.1860.0119.

G.P. Szegö. *Orthogonal Polynomials*. American Mathematical Society, Providence, RI, 1992. ISBN 978-0-821-81023-1.

H. Takenaka, T.Y. Nakajima, A. Higurashi, et al. Estimation of solar radiation using a neural network based on radiative transfer. *Journal of Geophysical Research*, 116:D08215, 2011. https://doi.org/10.1029/2009jd013337.

V. Twersky. On propagation in random media of discrete scatterers. *Proceedings of Symposia in Applied Mathematics*, 16:84–116, 1964.

E. Ustinov. Geophysical retrieval, forward models in remote sensing. In *Encyclopedia of Remote Sensing*, pages 241–247. Springer, New York, 2014.

H.C. van de Hulst. *Light Scattering by Small Particles*. Dover Publications, Inc., New York, 1957.

H.C. van de Hulst. *Multiple Light Scattering: Tables, Formulas and Applications*. Academic Press, New York, 1980.

L.J.P. van der Maaten, E.O. Postma, and H.J. van den Herik. *Dimensionality Reduction: A Comparative Review*. Technical Report TiCC-TR 2009-005. Tilburg University, 2009. URL http://lvdmaaten.github.io/publications/papers/TR_Dimensionality_Reduction_Review_2009.pdf.

R.F. van Oss and R.J.D. Spurr. Fast and accurate 4 and 6 stream linearized discrete ordinate radiative transfer models for ozone profile retrieval. *Journal of Quantitative Spectroscopy and Radiative Transfer*, 75(2):177–220, 2002. https://doi.org/10.1016/s0022-4073(01)00246-1.

R. West, D. Crisp, and L. Chen. Mapping transformations for broadband atmospheric radiation calculations. *Journal of Quantitative Spectroscopy and Radiative Transfer*, 43(3):191–199, 1990. https://doi.org/10.1016/0022-4073(90)90051-7.

W.J. Wiscombe and J.W. Evans. Exponential-sum fitting of radiative transmission functions. *Journal of Computational Physics*, 24(4):416–444, 1977. https://doi.org/10.1016/0021-9991(77)90031-6.

T. Wriedt. *Mie Theory: A Review*, pages 53–71. Springer, Berlin, Heidelberg, 2012. ISBN 9783642287381.

P. Yang and K.-N. Liou. *Light Scattering by Ice Crystals: Fundamentals and Applications*. Cambridge University Press, 2016.

P. Yang, H. Wei, H. Huang, et al. Scattering and absorption property database for nonspherical ice particles in the near- through far-infrared spectral region. *Applied Optics*, 44(26): 5512–5523, 2005. https://doi.org/10.1364/AO.44.005512.

G. Yanovitskij. *Light Scattering in Inhomogeneous Atmospheres*. Springer, Berlin, Heidelberg, 1997.

E.P. Zege and A.A. Kokhanovsky. Analytical solution to the optical transfer function of a scattering medium with large particles. *Applied Optics*, 33(27): 6547, 1994. https://doi.org/10.1364/ao.33.006547.

E.P. Zege, A.P. Ivanov, and I.L. Katsev. *Image Transfer Through a Scattering Medium*. Springer, Berlin, Heidelberg, 1991.

S. Zucker, D. Batenkov, and M.S. Rozenhaimer. Physics-informed neural networks for modeling atmospheric radiative transfer. *Journal of Quantitative Spectroscopy and Radiative Transfer*, 331: 109253, 2025. https://doi.org/10.1016/j.jqsrt.2024.109253.

Index

Analytical Methods in Radiative Transfer, First Edition. Alexander Kokhanovsky, Vijay Natraj and Dmitry Efremenko.